The Green Utility

The Green Utility

A Practical Guide to Sustainability

Cheryl Welch
Tualatin Valley Water District

American Water Works Association

The Green Utility

Disclaimer

This book is provided for informational purposes only, with the understanding that the publisher, editors, and authors are not thereby engaged in rendering engineering or other professional services. The authors, editors, and publisher make no claim as to the accuracy of the book's contents, or their applicability to any particular circumstance. The editors, authors, and publisher accept no liability to any person for the information or advice provided in this book or for loss or damages incurred by any person as a result of reliance on its contents. The reader is urged to consult with an appropriate licensed professional before taking any action or making any interpretation that is within the realm of a licensed professional practice.

AWWA Publications Manager: Gay Porter De Nileon
Acquisitions Editor: Mary Kay Kozyra
Cover Design/Production Editor: Cheryl Armstrong

Library of Congress Cataloging-in-Publication Data

Welch, Cheryl.

The green utility : a practical guide to sustainability / Cheryl Welch.

p. cm.

Includes bibliographical references and index.

ISBN 978-1-58321-792-4

1. Water utilities--Environmental aspects. 2. Water utilities--Management.
I. Title.

HD4456.W45 2010

363.6'10684--dc22

2010028100

Printed on 10% Recycled Paper

American Water Works Association

6666 West Quincy Avenue
Denver, CO 80235-3098
303.794.7711
www.awwa.org

Contents

List of Figures

Acknowledgments

This book exists only because of the input and assistance of many, many people. First and foremost I thank my partner, Beck, whose support, flexibility, and understanding of the time and effort it takes to write a book made this endeavor at all thinkable for me. I truly could not have done this without you.

Because I will never know all there is to know about sustainable water utilities, I called on several experts for assistance. Steve Conrad, of Simon Fraser University's School of Resource and Environmental Management, was especially generous with his time and knowledge. Steve and those listed below contributed valuable sections of this book, for which I am very grateful.

Stacey Barr, Performance Measure Specialist
Steve Carper, Tualatin Valley Water District
Kim Dinan, Portland Water Bureau
Collin Fleming, Tualatin Valley Water District
Dan Haddock, Layne Christensen Company
Mary Hansel, Carollo Engineers
Judi Hansen, Sydney Water
Doug Huxley, CH2M HILL, Inc.
Jim Meierotto, Tualatin Valley Water District
Francis Pamminger, Yarra Valley Water
Ryan Smith, Tualatin Valley Water District
Priya Sathyanarayan, CH2M HILL, Inc.

In addition, Dorothy Atwood, Bernice Bagnall, Bruce Barbarasch, Suzanne Chiavari, Cheryl Davis, Bagher Dayyani, Andrew DeGraca, Greg DiLoreto, Debbie Erickson, Dale Fishback, Paul Fleming, Judi Hansen, Regina Hauser, Darcy Hitchcock, Todd Heidgerken, Marcia Isbell, Becky Jones, Tony Kelly, Joan Kersnar, Mark Knudson, Brenda Lennox, Eva Lucia, Carol Maas, Jim Meierotto, Susan Millhauser, Bob Noelle, Jeremy Rann, Jim Ure, Susan Wade, Steve Whipp, Marsha Willard, Ray Yep, Doug Yoder, and many others added facts, figures, experiences, and knowledge that vastly improved what I was able to offer. I very much appreciate your willingness to share what you know.

Numerous people read drafts and made suggestions throughout the writing of this book, and I appreciate your valuable input as well. My thanks to Brad Bailey, Steve Conrad, Greg DiLoreto, Mary Kay Kozyra, Beck Lipp, Sue Parrish, Bob Rieck, Douglas Tsoi, Jim Ure, Marsha Willard, and Will Winn for that help.

The staff, management, and elected officials of the Tualatin Valley Water District have let go of traditional models, opened their minds to new ideas, and been creative beyond measure as sustainability has infiltrated the inner workings of this organization. Thank you for being such good role models and providing much of the substance of this book.

Thanks also to the American Water Works Association for understanding how important it is for water utilities to address sustainability issues as soon as possible and for making it easier for utilities to understand their options and act decisively.

Finally, to the incredible water utilities and specific staff members doing pioneering work in sustainability, many of whom you will meet in this book, my sincere admiration—you are inspiring to every one of us as we all learn more about how to live and work well within the bounds of nature.

It's the same each time with progress.
First they ignore you,
then they say you're mad,
then dangerous.
And then there's a pause and
you can't find anyone who disagrees with you.
Tony Benn, British politician

Prologue

I came to possess a "sustainability" mind-set from a variety of angles.

My parents grew up in the Depression era and were masters at saving, fixing, reusing, and jerry-rigging. It didn't matter whether or not we had enough money to go buy something new. If there was a way to keep using what we had, we did it. Throwing something in the trash was a last resort. Unnecessary waste was not an option.

Then there was Girl Scouts. The time I spent in that organization, especially at camp, helped me discover the natural world and its inhabitants. I became passionate about preserving that natural beauty and diversity. I just couldn't believe that humans were better than nature and had permission to use up whatever we wanted at the expense of other species. It didn't make sense.

And growing up, I also learned the concepts of stewardship and respect. I came to understand that we are all in this together and are all interdependent. I continue to believe that if everyone respected themselves, other human beings, other species, and the Earth, we would be in fine shape as a society. Not so easily done, it seems.

The notion of sustainability encompasses everything I believe in. It takes into account the economic, environmental, and social aspects of what is needed for the Earth and all of its inhabitants to survive far into the future. It acknowledges that the environment cannot be protected at all costs, just as the economy cannot take precedence over everything else. All must be balanced. It recognizes that we all need to work together to make that happen and that it will, indeed, take a lot of work.

About This Book

Water industry employees are charged with protecting perhaps the world's most valuable resource. In developed nations, we are so dependent on this resource that we tend to take it for granted, not affording it the value it deserves. As we look for ever more effective ways to manage our organizations so that we are able to provide this resource to all who need it, sustainability concepts will be of immense value.

In part, this book reflects Tualatin Valley Water District's path toward sustainability. As such, we hope others can learn from our journey. But it also reflects the inspirational experiences of many others. It is a collection of examples, industry references, and foundational concepts that goes beyond any one organizational perspective.

This book focuses specifically on water utilities. It shows what others in the industry have done to move their organizations along the path toward sustainability. It includes ideas, plans, and tools to make it as easy as possible to reduce negative effects on the environment, maximize positive impact in the community, and keep delivering water at a cost that reflects its value, but allows everyone to receive all they need. It is meant to inspire and encourage you to quickly take important steps in your own organization to do this.

While this guide is intended to provide a succinct overview of the many opportunities water utilities have available to become more sustainable, it is not a one-size-fits-all solution. Every organization will have its own personalities, infrastructure issues, financial position, political issues, energy costs, etc. Each agency will have people with different skills and interests that may make one opportunity more attractive than another. A suggestion that seems too simplistic for one person may seem out of reach to another.

There is no perfect path to sustainability. You will need to evaluate everything you find in this book to see if it suits your situation. Not everything will make sense for you. On the other hand, there are few fatal errors. Increasing your positive impact and decreasing your negative impact in any of myriad ways possible is going to be good. Anything you do is a start. Don't wait for the perfect plan to take your next (or first) step.

This book may be used as a reference guide to help you at whatever place you are in your journey. You don't need to read it from cover to cover, but if you are just starting out, it may give you a good road map to follow. The more I wrote, the more material I realized existed that I had not even touched on. Sustainability is so vast, even when only addressing water utility issues, that one volume could never hope to hold it all. At some point I had to stop in order to get the book out while it was still of some value. I hope that you will find what I have included helpful and be inspired to discover more on your own.

To maximize the usefulness of this book, look for the following aids:

TIPS: Tips are just things to think about. These will describe the benefits and challenges of a particular process, piece of equipment, or behavioral change. They may help you make the business case for selecting a more sustainable option. Or they might just be obstacles or hints others have discovered.

You will also find described some of the challenges you may face in implementing a process or changing equipment. These incorporate the experience of many water utility employees and may save you from going down a path not right for you.

Case Studies

People from all over the world have discovered or invented solutions for their own unique problems. They have found ways to be more efficient or use a safer product or process. They have found uses for what was once waste or collaborated in ways previously unheard of. The stories of many clever and resourceful people are included to instruct and inspire you. Maybe you can use their solutions as is. Or maybe their answers will spark a similar idea from someone in your organization that will be the perfect resolution to your specific needs.

Resources

This book is intentionally compact so you don't have to wade through more information than you require. But there are many wonderful books, articles, and tools available from AWWA, the Water Research Foundation, US Environmental Protection Agency, and others that may be important to your journey toward sustainability. Some of these are listed throughout the book to assist you in delving deeper into areas as needed.

Chapter 1

What Does Sustainability Mean to a Water Utility?

DEFINITIONS OF SUSTAINABILITY

So what is sustainability? One commonly accepted definition comes from the 1987 United Nations Brundtland Commission report that defines sustainability, or more specifically, sustainable development, as development that "meets the needs of the present without compromising the ability of future generations to meet their own needs."

Merriam-Webster's dictionary defines sustainability as "of, relating to, or being a method of harvesting or using a resource so that the resource is not depleted or permanently damaged." While these are excellent general definitions, and certainly applicable to the water industry, it may be even more helpful to see how various utilities customize these definitions in their public utility settings.

The Tualatin Valley Water District's (Beaverton, Ore.) sustainability mission is "to provide its community quality water and customer service in a manner that protects the earth's environment, supports a strong economy and promotes a robust community" (Tualatin Valley Water District, 2009).

Yarra Valley Water (Victoria, Australia) "recognizes that the well being of our community and the strength of our economy are dependent on the health of our environment. We strive to provide water and sewerage services within the carrying capacity of nature and will inspire others to do the same" (Yarra Valley Water, 2009).

Philip Green, CEO of the United Kingdom's (UK) United Utilities says, "Our aim is to provide the best value utility service to our customers in a way that is sustainable, environmentally responsible and reflects our wider social impact."

The San Francisco Public Utilities Commission defines sustainability as its "commitment to implementing a triple bottom line framework through which it will responsibly manage the resources under its care, protect public health, and balance its social and environmental responsibilities to the citizens and community, while providing cost effective services to its ratepayers" (San Francisco Public Utilities Commission, 2008).

The Sydney (Australia) Water Purpose: to deliver essential and sustainable water services for the benefit of the community (Sydney Water, 2009).

The San Francisco-Oakland, Calif., metro area is home to the East Bay Municipal Utility District. Their strategic plan approach "applies a holistic view and strives to minimize waste; conserve energy and natural resources; promote long-term economic viability; support safety and well-being for employees, communities, and customers; and benefit society" (East Bay Municipal Utility District, 2008).

Miami-Dade County, Fla., views sustainability in this way: "Simply put, sustainability refers to our ability to sustain ourselves, our community and the natural resources on which we depend (like clean air and water) for our day-to-day survival. Sustainability recognizes the need for us to move away from using resources in an inefficient and wasteful way and move toward a smarter way of doing business. It also recognizes the important links between economic prosperity, social well-being and environmental health" (Miami-Dade County, 2009).

And AWWA's Sustainability Ad Hoc Committee defines sustainability like this: "Sustainability means providing an adequate and reliable water supply of desired quality—now and for future generations—in a manner that integrates economic growth, environmental protection and social development" (AWWA Sustainability Ad Hoc Committee, 2009).

The concept of sustainability is not so difficult to understand. How do we operate our organizations to meet our customers' water requirements indefinitely (sustainably) while ensuring the systems in which that water exists continue indefinitely? Successful execution is the challenge. On one hand, since water utilities manage one of the world's most important natural resources, our swift success at becoming sustainable is vital. On the other hand, no water utility (or city or country) is yet fully sustainable. We all, however, have myriad opportunities to improve our processes and policies toward that goal.

The purpose of operating in a more sustainable manner is not to add more work, cost, and complexity to your organization. It is to encourage you to think farther into the future and farther into the world outside your front door. It is about understanding that your agency is not only part of the economy but also part of a much larger organization—society—that is in turn part of an even bigger system—the environment. These ideas are commonly represented in one of the following two ways.

In Figure 1.1, the three sectors intersect with a sustainable center. Where all systems are strong, the organization is likely to be sustainable. This image may be useful to apply to a program within your agency or even a particular purchase you are considering.

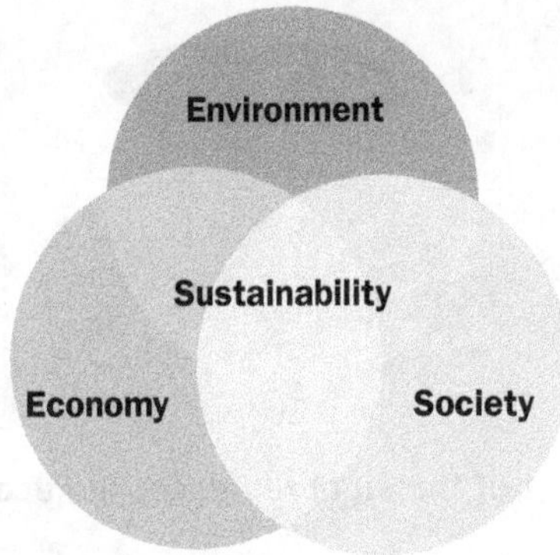

Figure 1.1. Sustainability as an intersection

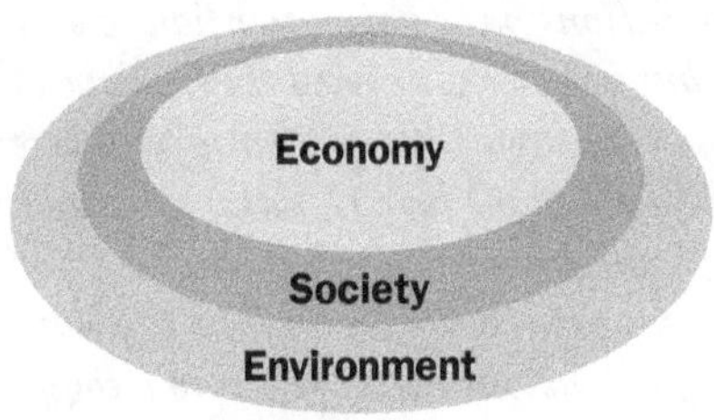

Figure 1.2. Sustainability as a nested set

A more practical, and probably more realistic, representation is shown in Figure 1.2. In this graphic, the economy is entirely encompassed by society, which is again fully enveloped by the environment. As Warren Buffet put it, "Society is a wholly owned subsidiary of the environment." For your organization to thrive, all of the systems in which you exist must be healthy.

It feels like an either/or decision sometimes to find the sustainable solution in any given situation. Paper or plastic bags. Electric hand dryers or paper towels. Disposable coffee cups or dishwasher. There are so many variables. But sustainability is less about finding perfect solutions and more about recognizing that organizational decisions concern more than just the economic bottom line. We need to consider the environmental and societal impacts of our operations, as well.

In trying to clarify that concept, many organizations often use another image—the three-legged stool (Figure 1.3). This analogy is useful for us in making decisions and explaining them to others. That is, that the three legs of economy, environment, and society must all be strong in order for the stool to stand. Visual aids and analogies like this are valuable in helping coworkers, management, elected officials, and the public comprehend this new way of looking at things.

Figure 1.3. The three legs of the stool represent the economy, the environment, and society

Some organizations have paid attention to each of the three legs for a long time, but haven't connected them in the context of sustainability. Almost all organizations have paid attention to at least one or two legs. So the point here is that most organizations are and have been doing something in a sustainable way for a long time. Everybody has some experience and success with one or more of the legs...now all organizations need to do is to recognize the other legs and pull it all together in an integrated way.

Bob Rieck, Portland (Ore.) Water Bureau, retired

And it is a new way to see. You will no doubt discover new opportunities just by looking at things through a sustainability "lens." Especially in older industries like ours, there are always opportunities for innovation. You look at a remote site and wonder why you have mowed it for all these years. You can't remember why you've never taken public transportation downtown, since you can get work done on the way. You look at a process and think, "What would a more natural, biological model of this look like?" Sustainability is about more than just eliminating waste. It will open up new ideas and opportunities you'd never thought of before.

A case in point is looking at how we design our infrastructure and operate our facilities. Biomimicry is a burgeoning field that suggests we look at how water moves through our pipes and tanks, what chemicals we add to purify it, how much nonrenewable energy we use, and more. It asks us to step back from the way we've always done things and look at what we might learn from nature. See the case study on the next page.

SUSTAINABILITY FRAMEWORKS

It is useful to have an underlying framework or foundation from which to engage in sustainability work. Tualatin Valley Water District (TVWD) chose The Natural Step (TNS) and the triple bottom line (TBL) to guide our development.

What if our built environment was fully sustainable? We know it is possible–forests, coral reefs, and other ecosystems all over the world have been doing it for millennia. Forests harness renewable energy to create materials and shelter, purify air and water, produce food and recycle all nutrients... it is a system performing the same functions that we need to perform... in a sustainable manner. If we look to the forest, or the coral reef, or the human circulatory system, what could we learn about sustainable design?

Now, imagine a municipal water system that:

- Maintained hydrological function of the watershed in which it is embedded;
- Sipped 100% renewable energy;
- Used only materials that you would feed your children for dinner;
- Eliminated unwanted by-products;
- Returned nutrients back to the soil;
- Created habitat.

This is the promise of biomimicry. Engineers and other water professionals are looking for ways to build more sustainable water systems. Some of those engineers are turning to nature for ideas. Biomimicry is an old practice and an emerging discipline that studies nature's best ideas and emulates them to solve human challenges. It is gaining momentum and driving sustainable innovation in a variety of sectors. It starts with the question, "How does nature...?"

An early example of applied biomimicry in the water sector is PAX Water's small, efficient mixer (Figure 1.4) that emulates nature's spiral flow geometry (Figure 1.5). When drinking water becomes stagnant and stratifies–especially in storage tanks exposed to strong summer sun–water in the upper layers can completely lose its disinfectant residual. This residual is needed to control bacterial regrowth and protect the health of consumers from waterborne illness. For many storage tanks, the solution to maintaining residual is to keep the water moving. Carollo Engineers worked with PAX Water to validate the bio-inspired Water Mixer for use in maintaining water quality through low-energy mixing in drinking water tanks and reservoirs. The Mixer can move the entire volume of a 7-million gallon tank using only 300 watts of power.

There are several other biomimetic products that could be explored for use in our industry. Examples include a nontoxic, antiscalant that mimics the "stop" proteins that regulate growth in seashells; a leak detection and repair product that mimics the platelets in our blood; and coatings to prevent biofouling that mimic signaling interference in seaweed and the rough surface of sharkskin. The coating based on sharkskin also reduces drag... could this improve flow through pipes to reduce pumping costs? Biomimetic wind and tidal power generator products are now on the market, and work is progressing on solar cells that capture solar energy similar to the way plants do through photosynthesis.

Specifying biomimetic products is a great place to start, but a more powerful way to apply biomimicry in our industry will be to create opportunities for biologists and engineers to collaborate in creating design solutions that solve problems on a whole-systems level rather than focusing on individual components. An example of this is the "ecological machine," engineered to clean wastewater by mimicking nature's strategies of diversity and self-selection to optimize treatment based on local conditions.

Figure 1.4. A bio-inspired water mixer is used to maintain water quality. Only 300 watts of power are used to move the entire volume of this 7-million gallon tank. (Photo courtesy of Carollo Engineers)

Figure 1.5. Calla lily in bloom (Photo courtesy of Free Nature Pictures)

Biomimicry offers a framework of "Life's Principles" to help facilitate approaching projects on the whole-systems level. All living organisms must adapt to the same operating conditions—earth is in a constant state of dynamic nonequilibrium, is subject to limits and boundaries, and is water based. A set of Life's Principles has been distilled from the biological literature to describe how organisms, species, and ecosystems sustain themselves in response to these operating conditions. Examples include being locally attuned and responsive, leveraging interdependence, recycling all materials, and developing resilient systems. Using the full list of Life's Principles to set project goals and as design and evaluation criteria can help ensure that a project delivers on sustainability objectives.

Applying biomimicry in the water sector is in its infancy. Serious sustainability challenges demand new solutions. We are surrounded by 30 million other species that have already solved the same challenges we face. Their strategies comprise a "sustainable solutions bank" that we can borrow ideas from to develop sustainable water systems that meet society's needs and protect the natural capital that holds the secrets to long-term survival. What would you ask nature?

Contributed by Mary L. Hansel, Carollo Engineers

Other possible frameworks are listed at the end of the section. Choose one of these or some combination that fits into the culture and politics of your situation. Or use a known framework as a starting place to develop something that makes sense to you and your stakeholders.

The Natural Step

This is a trademarked, science-based framework developed in Sweden by Karl-Henrik Robert to help organizations and communities understand sustainability and work to become more sustainable. It is also an international nonprofit organization based on this framework and dedicated to education, advisory work, and research in sustainable development.

The Natural Step framework's definition of sustainability includes four system conditions (scientific principles) that lead to a sustainable society. These conditions, which must be met to have a sustainable society, are listed below (reworded by SERA Architects, Portland, Ore.).

In a sustainable society, nature is not subject to systematically increasing:

- concentrations of substances extracted from the Earth's crust;
- concentrations of substances produced by society; or
- degradation by physical means.

And, in that society. . .

- people are not subject to conditions that systematically undermine their capacity to meet their needs.

In other words,

1. Can the earth replace what I take?
2. Am I poisoning the earth, water, or air?
3. Do I respect the biodiversity of flora and fauna?
4. Are the choices I make fair and equitable?

TVWD staff chose this as one of its two frameworks because we believed the science-based approach would appeal to the many engineers and environmental professionals among us. For the most part, it has done that, but there are still situations where TNS is a bit cumbersome and TBL (below) is more understandable. Some organizations use the TNS conditions to guide almost every decision they make. We tend to use TNS as an underlying set of principles but look to TBL when making specific decisions.

In The Natural Step framework, we saw many valuable characteristics that made this model useful to us. Besides appealing to the engineers, we also appreciated tools such as the funnel metaphor (Figure 1.6) for its simple but effective explanation of how "business as usual" will likely lead to environmental, social, or economic crises. This metaphor helped us visualize the economic, social, and environmental pressures that are growing on society as natural resources and ecosystem services diminish and the population's number and consumption grow.

The upper wall is the availability of resources and the ability of the ecosystem to continue to provide them. The lower wall is our demand for these

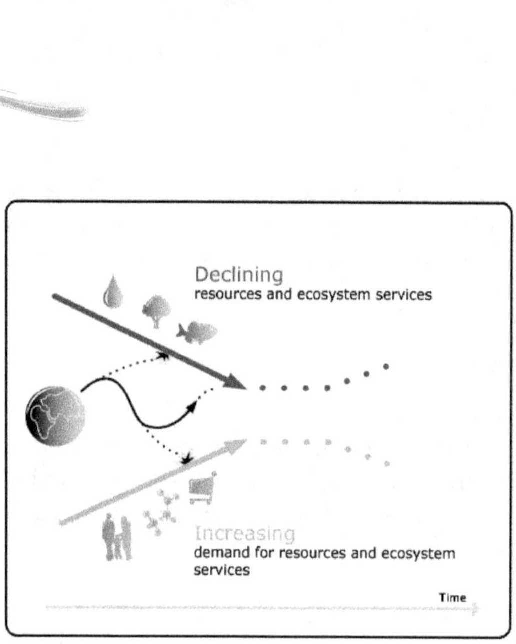

Figure 1.6. The Natural Step funnel of resources

resources, which are needed to make clothes, shelter, food, transportation, and other items, and the ecosystems that create them. It is easy to see that the closer the "walls" get, the less room we have to maneuver. This also shows, however, that reduced consumption and increased conservation of resources can open the funnel walls in the future.

The Natural Step contacts are available in Australia, Brazil, Canada, France, Italy, Japan, New Zealand, Singapore, South Africa, Sweden, Netherlands, the United Kingdom, and the United States. If you are located near others interested in The Natural Step, be sure to take advantage of its networking aspects. Corresponding and collaborating with other innovative organizations in different sectors can enrich your program and help you accomplish even more than you could do on your own.

Triple Bottom Line

The phrase, *triple bottom line*, was coined by John Elkington, cofounder of the business consultancy SustainAbility, in his 1998 book *Cannibals with Forks: the Triple Bottom Line of 21st Century Business.* This view of business and organizational accounting adds social and environmental values to the traditional economic measures of success. You might hear a similar concept described by such phrases as *People, Planet, Profit; Environment, Economy, and (Social) Equity;* or *Human Capital, Natural Capital, Manufactured Capital, and Financial Capital* (Hawken et al., 1999).

This concept is widely used, easily understood, and is something that can be adapted for any organization. As our utility began its sustainability work, our staff recognized that we, like most water utilities, were familiar with watching the economic bottom line, but less used to taking into account environmental and societal costs and benefits. As we examined our operations, however, we discovered we were already considering all three "bottom lines" more than we knew.

Much of the credit for this goes to a forward-thinking board, the Sustainability Steering Committee, we formed to review sustainability policy issues, and the creativity of our employees. Even prior to adopting The Natural Step framework and forming a Sustainability Steering Committee, the District had initiated reductions of material waste and energy use. Management had ensured financial sustainability by preparing 50-year financial forecasts that were updated annually. The organization emphasized efficiency by reducing water loss and properly maintaining its operational systems. These are the kinds of actions you, too, can point to as the initial kernels of your sustainability program.

Tualatin Valley Water District, for instance, already had a long history of participating in the community. Since 2002, we had assigned money to a Washington County nonprofit organization to screen and provide district customers with water bill financial relief when emergencies arose for these customers. Three acres of land adjoining district headquarters property had been loaned to the local parks and recreation district for use as sports fields until we had need of it. We shared our fuel tanks with the same agency, eliminating the need for them to install a separate tank, and keeping costs down for both organizations. They chipped up some of our wood debris for use on local trails—a beneficial partnership for both groups. A team of employees had been picking up litter along neighborhood streets for years. All these activities contributed to the social benefits we provided to our community.

Environmentally, we had been officially urging the conservation of water since 1993, when a conservation program was developed. That program was, and is, dedicated to reducing excess customer water usage through rebates on water-efficient fixtures and appliances; give-away items, such as faucet aerators, shower timers, and leak kits; and supplying free advice in the form of educational publications, assessments, and workshops. In 1993, the District constructed its first demonstration garden showing customers how native plants and grasses can make a difference in their water use. Conservation program staff is very active with the area's Regional Water Providers' Consortium and can always be found at local and regional events, touting the benefits of sustainability and water conservation, in particular.

Water conservation is a critical topic unto itself. The American Water Works Association and other groups have many publications devoted strictly to this subject. For a wide selection of books and articles, start by logging on to AWWA's Web site, accessing the Water Library, and search under "Conservation" for more information.

In addition to conservation, the District was engaged in other programs. Recycling had been introduced, transit passes were given to each employee to encourage the use of public transit, and the local electric utility had been engaged to help replace our outdoor lighting with more efficient lamps to reduce energy usage and therefore electric bills. We had operated on a 9/80 work schedule (9-hour days for the most part, with every other Friday off) for many years, thereby allowing employees to eliminate commuting one day every two weeks. We had also adjusted work hours to a 7:00 a.m. to 4:30 p.m. schedule. That helped keep employees off the roads at the heaviest commuting times.

All this and more had been done before our employees ever used the word *sustainability*. We needed to find a framework that fit easily into what employees and management were already used to. A triple bottom line structure made sense to us. We could fit what had been done into each "bottom line" or leg of the stool and build new initiatives from there.

Look at your operations to see where you are already excelling. Listen to the phrases your work culture already incorporates, or the initiatives you already employ, perhaps 6 Sigma, Total Quality Management, Balanced Scorecard, or Lean Management. One sustainability framework or another may fit better with what employees already understand and are familiar with. And you may be surprised at the many actions you are taking that will help show you what framework will suit your needs.

Alternatively, you could take pieces of any of these frameworks or others and create your own framework or principles. It is just helpful to have a set of agreed on standards to refer back to when you need them.

Once you have established how you will view sustainability, then how will you know if you are making progress? There are many tools to help you assess where you are and where you want to go. But first let's look at the benefits a water utility might realize from operating in a sustainable manner.

Resources

AWWA Publications

- AWWA Water Library, http://apps.awwa.org/WaterLibrary/Search.aspx

Biomimicry

- Benyus, Janine. 2002. *Biomimicry: Innovation Inspired by Nature.* New York: Harper Perennial.
- Free, online database of nature's solutions that is searchable by design function at *www.asknature.org*. Ask "How does nature desalinate?", "How does nature filter?", etc.

Additional Sustainability Frameworks

- The Natural Step, www.naturalstep.org, http://www.naturalstep.org/en/the-funnel
- Ceres Principles, www.ceres.org/Page.aspx?pid=416
- Earth Charter, www.earthcharterinaction.org
- Hannover Principles, www.mcdonough.com/principles.pdf

References

AWWA Sustainability Ad Hoc Committee. 2009. Personal communication with Andrew DeGraca. 6/26/09.

East Bay Municipal Utility District. 2008. *Strategic Plan, Sustainability Principles.* http://ebmud.com/about-ebmud/strategic-plan. Accessed 1/10/10.

Hawken, P., A.B. Lovins, and L.H. Lovins. 1999. *Natural Capitalism.* http://www.natcap.org. Accessed 1/10/10.

Miami-Dade County. 2009. *Sustainability.* http://www.miamidade.gov/oos/. Accessed 1/10/10.

San Francisco Public Utilities Commission. 2008. *Sustainability Plan and Program 2008, Executive Summary.* http://sfwater.org/msc_main.cfm/MC_ID/18/MSC_ID/121. Accessed 1/10/10.

Sydney Water. 2009. *Annual Report, Sustainability Overview.* http://www.sydneywater.com.au/AnnualReport/sustainability/sustainability_overview.html. Accessed 1/10/10.

Tualatin Valley Water District. 2009. *Practicing Sustainability.* http://www.tvwd.org/about-us/practicing-sustainability.aspx. Accessed 1/10/10.

Yarra Valley Water. 2009. *Environmental Policy.* http://www.yvw.com.au/yvw/Home/AboutUs/OurPolicies/environmental-policy.htm. Accessed 1/10/10.

Chapter 2

Why Would a Water Utility Want to be More Sustainable?

Sustainability is not a fad; it is one of the emerging organizational imperatives for the 21st century. The general practices that fueled success and progress in the last century are no longer feasible given a world of exploding population and diminishing natural resources. As organizations that are assigned the responsibility of managing and providing one of the planet's most precious resources, water utilities have special motivation to adhere to the principles of sustainability.

The best retort to this chapter's title question is: Why would a water utility want to be UNsustainable? We want to ensure that we are providing enough quality water to satisfy our customers' needs well into the future. What other path would there be?

Some organizations, however, will have a harder time than others convincing elected officials, management, coworkers, and sometimes the public that sustainable actions are best for the organization's long-term health and accomplishment of its mission.

This chapter, therefore, outlines the most compelling reasons why sustainability simply makes good business sense to water utilities. Some relate to risks we can avoid, others to opportunities to improve our organizations' and communities' social, environmental, and financial positions. Put together, they should leave you no doubt that you are headed in precisely the right direction when you pursue a more sustainable organization.

Benefit to the Environment

Resource Management

Environmental stewardship, as defined on the US Environmental Protection Agency's (USEPA) Web site, is "the responsibility for environmental quality shared by all those whose actions affect the environment" (USEPA, 2009b). Whether it is important to you and your colleagues to enjoy recreational activities in natural settings, to leave a livable world for future generations, or to support healthy air, water, and other natural resources, we all have reasons to pay attention to how our actions affect the environment and our communities.

More importantly, resource management is the very business we are in. Our utilities have been charged with the protection and, indeed, sustainability of water. We, of all people, should understand the importance of using resources wisely and developing and maintaining safeguards for those resources.

This is the best argument in favor of sustainability. If we don't use all of our resources as efficiently as possible, we can't fulfill our mission into the future in the best way possible. We are the caretakers for future populations and have signed on to be responsible stewards.

The opportunities for new products, services, and ways of looking at our current systems are actually quite exciting. But we get used to doing things in a certain way and it is often hard to see change as an exciting new possibility instead of an imposition. The challenge is to frame our stewardship and conservation of resources as a chance to develop products or systems even more useful than those we now have.

Climate Change Issues

The Intergovernmental Panel on Climate Change (IPCC) was formed in 1988 by the United Nations (UN) Environmental Programme (UNEP) and World Meteorological Organization (WMO) to provide objective information about climate change to the public and to policymakers. IPCC is made up of scientists and experts from all over the world and promotes the UN goals of human development.

In early 2007, IPCC published a report that stated that there will be serious consequences unless people act to reverse climate change: rising sea levels; more frequent and less predictable floods (that will put millions of people in low-lying regions at risk); severe droughts; famine, particularly in Africa and central Asia; and the loss of up to a third of our plant and animal species. Diseases such as malaria and dengue fever will be spread more easily.

According to IPCC, "climate change refers to a change in the state of the climate that can be identified (e.g., using statistical tests)...by changes that persist for an extended period, usually decades or longer. It refers to any change in climate over time, whether due to natural variability or as a result of human activity" (IPCC, 2007).

AWWA and the Water Research Foundation (WRF, formerly AwwaRF) both have experts examining how changes in our climate may affect water utilities throughout the world. In AwwaRF's 2006 publication, *Climate Change and Water Resources: A Primer for Municipal Water Providers*, the following are listed as issues specifically relating to water utilities:

- Water utilities are vulnerable to climate-related disruptions even in the absence of climate change

- The Earth's climate is changing and will continue to change in coming decades
- Warmer temperatures have already had significant hydrologic impacts
- Drought-related wildfires can increase a utility's vulnerability to water quality impairment and loss of reservoir storage capacity as a result of soil erosion and debris flows

The authors of the AwwaRF publication summarize the effect of climate change as follows:

> Scientists agree on some of the important broad-scale features of the expected hydrologic changes, the most likely of which will be an increase in global average precipitation and evaporation as a direct consequence of warmer temperatures....
>
> At the regional scale, precipitation predictions are less certain. Changes in circulation patterns will be critically important in determining future changes in precipitation and water availability, and climate models can provide only a crude picture of how those patterns may change. The currently available evidence suggests that arctic and equatorial regions may have a tendency to become wetter, and that subtropical regions may experience drying. Projections of precipitation changes for temperate regions are less consistent.
>
> The water supply for any utility will depend on the quantity and timing of local and regional precipitation, both of which may change with global climate change. While it is impossible to make reliable predictions of change in the overall quantity of precipitation for a particular region, scientific theory suggests an intensification of the global hydrological cycle, leading to more intense but possibly less frequent periods of precipitation.
>
> In other words, we may see longer periods of drought alternating with spells of heavy rainfall and runoff. Such changes could create a number of difficulties for water utility planning and operations (Miller and Yates, 2006).

Carbon Emissions Regulation

Climate change is largely the result of carbon emissions given off by the burning of carbon-based fuels. Many of the gases emitted by the use of fossil fuels contribute to the warming of Earth's atmosphere. These gases are often called *greenhouse gases* (GHGs), after the greenhouse effect, discovered by Joseph Fourier in 1824 (Greenhouse Effect, 2010). The greenhouse effect is the heating of the surface of a planet or moon due to the presence of an atmosphere containing gases that absorb and emit infrared radiation. In the water industry, the greenhouse gases your operations are most likely to emit are carbon dioxide (CO_2), methane (CH_4), and nitrous oxide (N_2O).

USEPA has now declared that GHGs threaten the public health and welfare of the American people (USEPA, 2009a). Coming regulation of these gases may be reason enough to take a good look at this issue. Customer and community

Responding to Climate Change in Sydney, Australia

For an urban water utility such as Sydney Water, some of the key areas of interest in climate change include improved prediction of changes in water supply and customers' water demand caused by hotter climate and lower rainfall. Sydney Water also aims to better understand changes in the frequency, intensity, and length of extreme events, such as heat waves, bushfires, floods, storms, and surges in sea levels that might result from increased storm intensity.

During 2007-08, Sydney Water, in partnership with the CSIRO (Commonwealth Scientific and Industrial Research Organisation), the University of New South Wales (NSW), and other government agencies, took part in ongoing research to better understand and address the effects of climate change and variability on its assets and services. The CSIRO predicts that in around 2030, the average yearly rainfall for Sydney may decrease by 3% (relative to 1990, for a midrange emissions scenario). Rainfall decreases may be particularly large in the winter and spring months. Evaporation is also expected to increase.

To respond to these challenges, Sydney Water has developed a Climate Change Strategy and Action Plan to effectively identify and address the risks associated with climate change.

The plan outlines how Sydney Water will:

- Research... to better understand the extent of its carbon footprint and the impacts of climate change;
- Mitigate... by cutting its emissions by 60% by 2012, becoming carbon neutral by 2020, and by engaging with business and the community to reduce emissions; and
- Adapt... to reduce the risk of adverse impacts on operations and infrastructure, and improve the sustainability of freshwater supplies.

Sydney Water is one of the largest energy users in NSW. Pumping stations and treatment plants use a lot of electricity to treat and move water and wastewater around. Sydney Water aims to make the best use of energy and electricity and reduce its carbon footprint. Sydney Water aims to reduce emissions by 60% by 2012 and be 100% carbon neutral for energy and electricity use by 2020. Sydney Water will reduce its emissions from energy and electricity by reducing energy use, being more energy efficient, and using carbon offsets. This will ultimately eliminate or offset over 400,000 tonnes of carbon dioxide a year. This is the same as taking 100,000 cars off the road.

See http://www.sydneywater.com.au/Sustainability/carbonneutral.cfm for more information.

In April 2008, as part of this strategy, Sydney Water completed a qualitative Climate Change Infrastructure and Operations Risk Assessment. This involved conducting a workshop-based risk assessment consistent with the process outlined by the Australian Greenhouse Office (now Department of Climate Change) (2006). The risk assessment confirmed that Sydney Water is well positioned to address climate change risks as we deal with weather-related events on a daily basis and have a thorough understanding of how the climate influences our business from over a century of experience in dealing with extreme natural events. We now need to plan for the fact that climate change has the potential to alter the likelihood and consequence of those risks occurring. Sydney Water will use the results of this assessment to prioritize the research efforts. The focus of the next two to three years will be in developing appropriate tools to improve our understanding and assessment of climate change risk.

For further information, see Sydney Water's Risk Assessment in assessing climate change.

pressure, as well as general efficiency and stewardship of resources, will also be important drivers.

Fossil Fuel Availability

Petroleum reserves are finite. What is not known, and what is hotly debated, is the timing, severity, and duration of the coming decrease in oil production. As was mentioned in the Resource Management section, even without this knowledge, it is likely in our best interests to find ways to use less of this resource, making it last as long as possible. Fuel prices have been extremely volatile for many years, posing significant budgeting and operational challenges.

In the water industry, much of the oil we use comes in the form of petroleum-based electricity. We use electricity for pumping and for heating, cooling, and lighting our facilities. Petroleum is used for our construction, maintenance, and general fleet needs and even to get our employees to work in the first place. Think about how you would accomplish your mission if fossil fuels were unavailable, or if they were many times the price they are now—could you do it?

Waste Reduction

There is just no "away" when you throw things away. It will not come as a surprise that just because your garbage is not within your physical confines does not mean it has been handled in a manner that is beneficial, or at least not harmful, to the environment.

Waste takes on many forms, including:

- waste of natural resources,
- waste of human potential or energy,
- waste of transportation of products,
- waste of products that have no further use,
- financial waste, and
- waste of by-products that could be used as feedstock elsewhere, but aren't being so used.

Customers want to know you are operating efficiently with their rate or tax dollars. Waste is not efficient. More than that, however, what we now consider waste has potential to be used for other products. So you may not only reduce disposal costs, you might find an actual source of income for what you once threw away. Our "waste" is far more valuable in new jobs and products created when we find another use for it than when it is buried in a landfill.

And the issue is not just increasing recycling or reuse. We have an even more important opportunity to go back further in the process that creates waste and move from a linear "take-make-waste" mind-set to a circular "borrow-make-use-return" model of resource usage. This is sometimes termed *cradle-to-cradle* (Braungart et al., 2002) or *zero waste* (Zero Waste Alliance, 2009)—ideas that take their inspiration from nature, in which there is no place for the concept of waste, but in which waste = food. It may surprise you what is being thrown away by your utility that could be used by someone else—or what you could use that is waste from another.

Benefit to the Utility

Risk Management

The financial community is paying attention to sustainability. Swiss Re has said that if a company does not take its carbon footprint seriously, maybe their company is not interested in insuring it. JPMorgan Chase has asked clients who are large emitters of GHG to develop reduction plans, following similar commitments by Citigroup Inc. and Bank of America Corp (Lovins, 2006). Sarbanes Oxley maintains, in part, that if you fail to disclose to shareholders anything that can materially affect share price, you can be held personally criminally liable. Increasing numbers of stakeholders are interested in how you operate, and sustainable practices can actually calm questioners' fears in many cases.

Reduced Operating Costs

Sustainable innovation is often simply cost effective, making it ever so easy to do the "right" thing. Energy and water efficiencies, along with waste-reduction strategies, will frequently cut costs and save money. First, investigate what inefficiencies are occurring. Some of the sections in chapter 3 will assist you with that.

Anything from faucet aerators to variable frequency drives can help eliminate excess use of resources. With additional outlay, rainwater harvesting, solar panels, or wind turbines might be a good financial investment in many situations. Sustainable practices can often be easily justified with financial arguments.

Many of the almost 200 practices outlined in chapter 4 will result in economic savings in addition to environmental and community benefits.

Demand-side Management

The cheapest unit of water is the one we don't have to produce. On one hand, we all need jobs and we need the income from our customers to keep our utilities

The Tualatin Valley Water District (TVWD) was fortunate enough to have commissioners and management who believed a sustainable water utility was important to the viability of the community and the survival of the environment, in addition to our overall success as a water provider.

The driving forces behind our organization's decision to etch sustainability into all facets of our operations ranged from political to ecological. Some of our considerations:

- The effects of climate change on the Tualatin River and Bull Run watersheds are likely to mean shorter, more intense periods of rainfall and longer dry periods during the critical summer months.
- Decreasing supply could cause serious constraints for our customers if the Portland metro area continues to grow by 1 million people by 2030, as predicted.
- This expanding population is expected to create a 50% increase in water demand by 2050.
- Dwindling salmon populations from reduced river and stream flows are causing alarm and creating greater restrictions on water withdrawals.
- Diminishing fossil fuel supplies and geopolitical insecurities in oil supply, along with the climatic threat of carbon emissions, encourage us to plan for future scenarios with fewer fossil fuel options.
- As technologies for measuring contamination become more advanced, regulation is likely to become more stringent, and we recognize our responsibility to communicate this information openly to the public.
- With rising costs of living and economic hard times, it is our challenge to sustainably manage a constrained resource while providing an affordable public good.

operating. On the other hand, the more we reduce consumption of water, the less infrastructure we are required to add and maintain, reducing costs to our customers. Encouraging customers to be efficient (sustainable) in their usage of our product can be an economic benefit to us in the long run, potentially deferring bond issuances and large capital projects, and allowing more time for planning.

Asset Management

Asset management is all about sustainability and sustainability is all about asset management. This field of study advises us to maintain all of our assets in the most efficient manner possible. This applies not only to infrastructure but to money, people, and natural resources as well. It guides the addition, use, and disposal of assets in order to make the most of them throughout their lives. And it manages and maintains the costs and risks associated with the assets.

Sustainability also gives us another set of lenses through which to view asset management. We see different things when we put on "sustainability glasses." Asset management traditionally covered a utility's basic infrastructure. By understanding that people and the environment are also assets, we add new layers to this undertaking. And when we consider all of the layers in our decisions, we gain the opportunity to make better, longer-lasting choices.

Staff Health, Morale, and Productivity

People like working in situations where they feel respected and taken care of. Many people already have sustainable practices built into their own lives and appreciate those opportunities at work. Introduction of similar practices into the workplace will generally be received positively since staff want to work for employers they can be proud of. Employee turnover is expensive, and being socially and environmentally responsible may gain you as much as high salaries with some candidates.

Sustainable buildings have been shown to be good for staff morale, increase staff productivity, and reduce absenteeism, as well as provide a clean, healthy environment for customers and visitors. L. Hunter Lovins, in a 2006 University of Washington speech, reports one Boeing lighting system retrofit that cut lighting energy costs by 90%, and because workers could also see better, it reduced their error rate by 30%—a win for the company, its staff, and ultimately Boeing's customers (Lovins, 2006).

Public Relations

Talking to customers about sustainability may stimulate a deeper understanding of water as a resource and of your utility in particular. Many people take for granted that quality water will come out of their faucets whenever they need it for as long as they need it. Reaching customers and community members who are passionate about sustainability may give you a receptive audience to educate about the complexities of getting their water to them—an audience that may never have thought about these issues before.

Sustainability can get you noticed by the media, too. It is always good to have a positive reason to be in the paper or on the Web. A new sustainability-related project can be the perfect opportunity for a press release that will gain customer and community respect for your utility. Be careful, though, of tooting your own horn too loudly. Stories can backfire if you haven't done your homework in some area. It is often best to be public about your accomplishments, but humble in your characterization of them.

Benefit to Society

Reduced Negative Impact Within the Community

Reduced resource usage—electricity, paper, building materials, and water—leaves more for other citizens now and for future generations. Less pollution, in the form of carbon emissions, solid waste, light spilling into neighboring houses at night, or other negative contaminants, will improve the quality of life for people far beyond your physical borders. Fewer toxic pesticides and other chemicals reduce the risk of contamination of the communities in which you operate. This improves your community standing, leading to the next benefit.

Increased Positive Impact Within the Community

Being a better "corporate citizen" is a natural result of increasing your organization's sustainable actions and policies. Your utility will be helping to create a more healthy society—environmentally, financially, and socially—as you consider all three of these "bottom lines" in your daily decisions. Developing good relationships within your community builds a solid foundation and creates respect for your organization. When other issues arise, from main breaks to bond issuance, you have credit that can be worth a lot to your utility.

The highest leverage place to think about waste is upstream. Waste prevention, i.e., reducing consumption (and therefore reducing production of additional stuff), far outweighs—by orders of magnitude—the environmental and carbon footprint of recycling and reusing on the downstream end. The "environmental cost" of waste disposal is typically less than 5% of the total environmental cost that includes production and use. Recycling reduces upstream impacts. Prevention (not consuming) eliminates upstream impacts.

Achieving "true" or full zero waste needs to be done as part of a system. Think out of your organization walls and look to collaboration to take your zero waste efforts to the next level.

Dorothy Atwood,
Zero Waste Alliance

Collaborative Opportunities

One trademark of sustainable thinking is its holistic quality. Sustainability is about making whole systems work for all involved. Look for opportunities with other water agencies or local community organizations to combine your efforts or share your resources to increase the efficiency and effectiveness of your operations. From sharing resources (tools, construction equipment, meter-reading services) to developing an integrated watershed plan or combining forces on regional conservation advertising, there are many ways to combine forces. These efforts may save you money, time, or carbon emissions, and it is worth the time to look around outside of your own organization every so often to see what opportunities exist for cooperation.

Enhanced Ability to Plan for the Future

Our customers expect water to always be available in the quantity and quality that they want it. Reduction of risks, reduction of exposure to future regulation, increased efficiency, and improved asset management all lead to a utility's ability to better plan for the future. This, in turn, enables us to better serve the public's needs and maintain or improve the standard of living for its citizens.

Seen together, the above should represent a compelling argument for a sustainable water utility philosophy. Now, let's get into how to make it happen!

Resources

Asset Management and Sustainability

- Marlow, D., D. Beale, and S. Burn. 2010. *Linking asset management with sustainability: Views from the Australian sector. Jour. AWWA*, January 2010. http://www.awwa.org/publications/AWWAJournalArticle.cfm?itemnumber=53235.

Carbon Footprint Resources

- Griffiths-Sattenspiel, B. and W. Wilson. 2009. *The Carbon Footprint of Water*. River Network. May 2009.
- Miller, K., D. Yates, C. Roesch, and D.J. Stewart. 2006. *Climate Change and Water Resources: A Primer for Municipal Water Providers*. UCAR and AwwaRF.

References

Braungart, M., and W. McDonough. 2002. *Cradle to Cradle: Remaking the Way We Make Things*.

Greenhouse Effect. In Wikipedia, the free encyclopedia. http://en.wikipedia.org/wiki/Greenhouse_effect. Accessed 10/2/09.

IPCC (International Panel on Climate Change). 2007. Fourth Assessment Report, Section 9.1.1. Cambridge, UK: Cambridge University Press. http://www.ipcc.ch/publications_and_data/ar4/wg1/en/ch9s9-1.html.

Lovins, L.H. 2006. http://uwtv.org/programs/displayevent.aspx?rID=9320. Program on the Environment talk at University of Washington 11/8/06.

Miller, K., and D. Yates. 2006. *Climate Change and Water Resources: a Primer for Municipal Water Providers*. Denver, Colo.: Water Research Foundation.

USEPA. 2009a. http://www.epa.gov/aging/press/epanews/2009/2009_1207_1.htm. Accessed 1/10/10.

USEPA. 2009b. http://www.epa.gov/stewardship/. Accessed 9/12/09.

Zero Waste Alliance. 2009. http://www.zerowaste.org/case.htm. Accessed 9/12/09.

Chapter 3

Where Do We Start?

Figuring out what you do and how you measure your progress toward sustainability is one of those areas that can either inspire or immobilize a person or an organization. It is preferable to have baseline information about your usage of electricity, water, and other resources and to know your contribution to pollution and solid waste. But don't let not knowing every bit of the preliminary data be an excuse for not moving forward. Much of what you can do to improve the sustainability of your utility will be clear to you without preliminary studies.

On the other hand, knowing where you are today shows you where you have room to improve. Understanding your current practices helps you figure out when and how new actions change your negative impacts on the environment and your positive impact in your community. Documenting and analyzing various inputs and outputs allows you to chart your progress. And where you are able to be systematic about your work, you will prosper by having data to show upper management, elected officials, and the public about where you've been and how far you've come.

As Lewis Carroll said, "If you don't know where you are going, any road will get you there." Balance your efforts between making visible progress that employees, management, and elected officials can be proud of and envisioning where you'd really like your actions to take you.

Strategic Sustainability Planning

Strategic planning is a management tool that helps an organization focus on the future and ensures that everyone is working toward the same goals.

Strategic planning generally involves the assessment of your organization's current position, the identification of where you'd like to be, an evaluation of issues to be resolved to get there, and a plan to make it happen. A sustainability plan follows much the same course. If you already do strategic planning, there is likely no reason to do a separate sustainability plan. You can expand your strategic planning process to include the elements of a sustainability plan.

A sustainable decision is one that weighs up all relevant social, environmental, economic and technical considerations in a way that is understood and accepted by key stakeholders.

Sydney Water Sustainability Planning Framework

If you do not already engage in a formal strategic planning process, however, developing a sustainability plan could be perfect for your needs. One book, *The Step-by-Step Guide to Sustainability Planning* (Hitchcock et al., 2008), is a good place to start. The authors' description of the steps for creating a sustainability plan is shown in Figure 3.1. Most of these steps are addressed in this book; all of them are addressed in theirs.

The electronic tool this book accompanies is called SPaRK™—the Sustainability Planning and Reporting Kit. It is a process for helping organizations quickly develop a sustainability plan and report format, with electronic files that help support that effort. The electronic files include an Excel® workbook with many of the common analyses and metrics an organization would use. This is linked to a Word® template for an internal sustainability plan and report so that the charts and graphs from the Excel file automatically show up in the right places. This is a good resource if you need a structured way to go about writing a plan.

The San Francisco Public Utilities Commission (SFPUC) has an excellent sustainability planning process that is detailed on their Web site under their Sustainability category.

Figure 3.2 is a visual representation of SFPUC's sustainability framework. It displays 6 categories, 22 subcategories, and 62 issue areas that frame their 163 indicators. Most organizations will not need, nor be able to effectively track, that many issue areas or indicators. But this diagram will give you a good idea of what you might choose from and how you might categorize what you do want to follow.

Sustainability Status and Progress

SCORE Assessment

One recommended instrument to monitor how well sustainability is integrated into organizational operations is the Sustainability Competency and Opportunity Rating and Evaluation (SCORE) assessment.

One way to use this tool is to evaluate your current sustainability status to see where opportunities exist to progress. The tool actually includes a "pre-SCORE," an assessment for organizations that are so new to sustainability that they want to use SCORE to understand what sustainability means and build a plan for when to address the practices instead of rating themselves on the assessment.

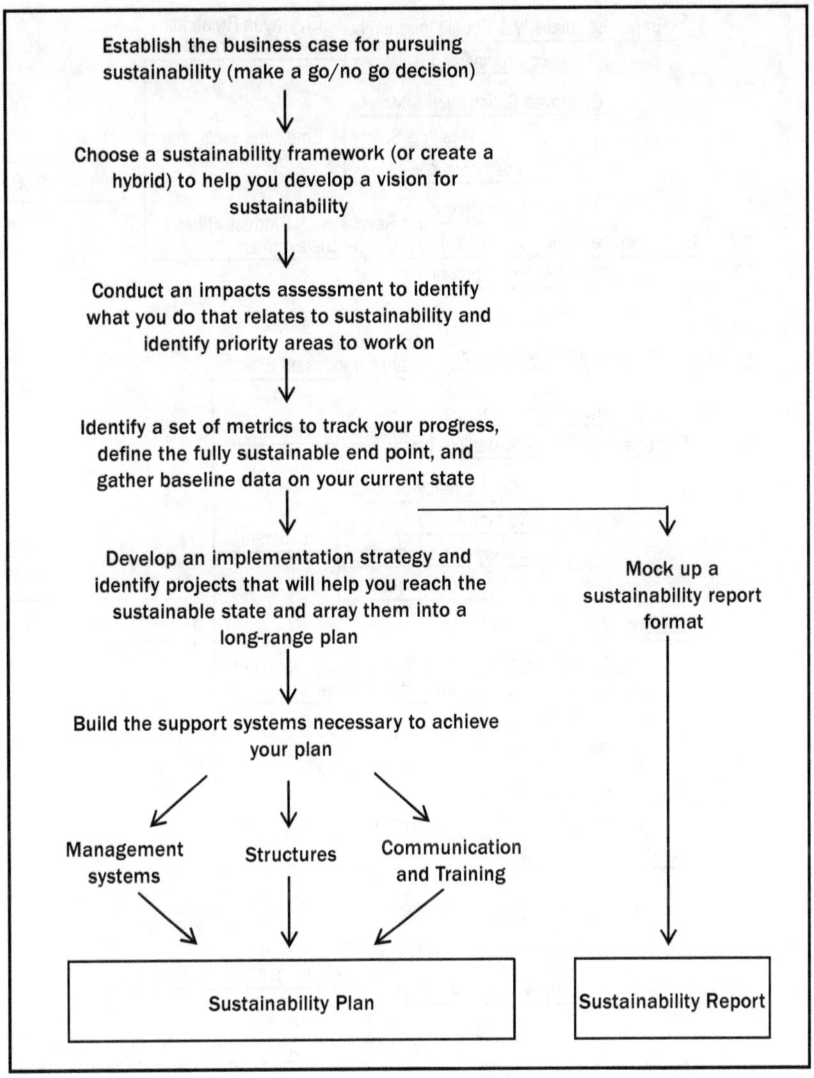

Figure 3.1. Steps for creating a sustainability plan

SCORE includes over 100 practices and is organized by functions that practically all organizations have (human resources, facilities, purchasing, finance, etc.). This makes it easy to assign responsibility and it helps departments understand what they are supposed to do.

For each practice, SCORE presents three levels of performance so you can see where you are now and where you need to go long term. The top level of performance is designed to be difficult to achieve. It describes the ultimate in sustainability for each particular practice. In most cases, none of us will be there yet. But it gives us a "North Star" to guide our planning.

In addition to the organizational functions, there are sector-specific assessments, including one for the water industry.

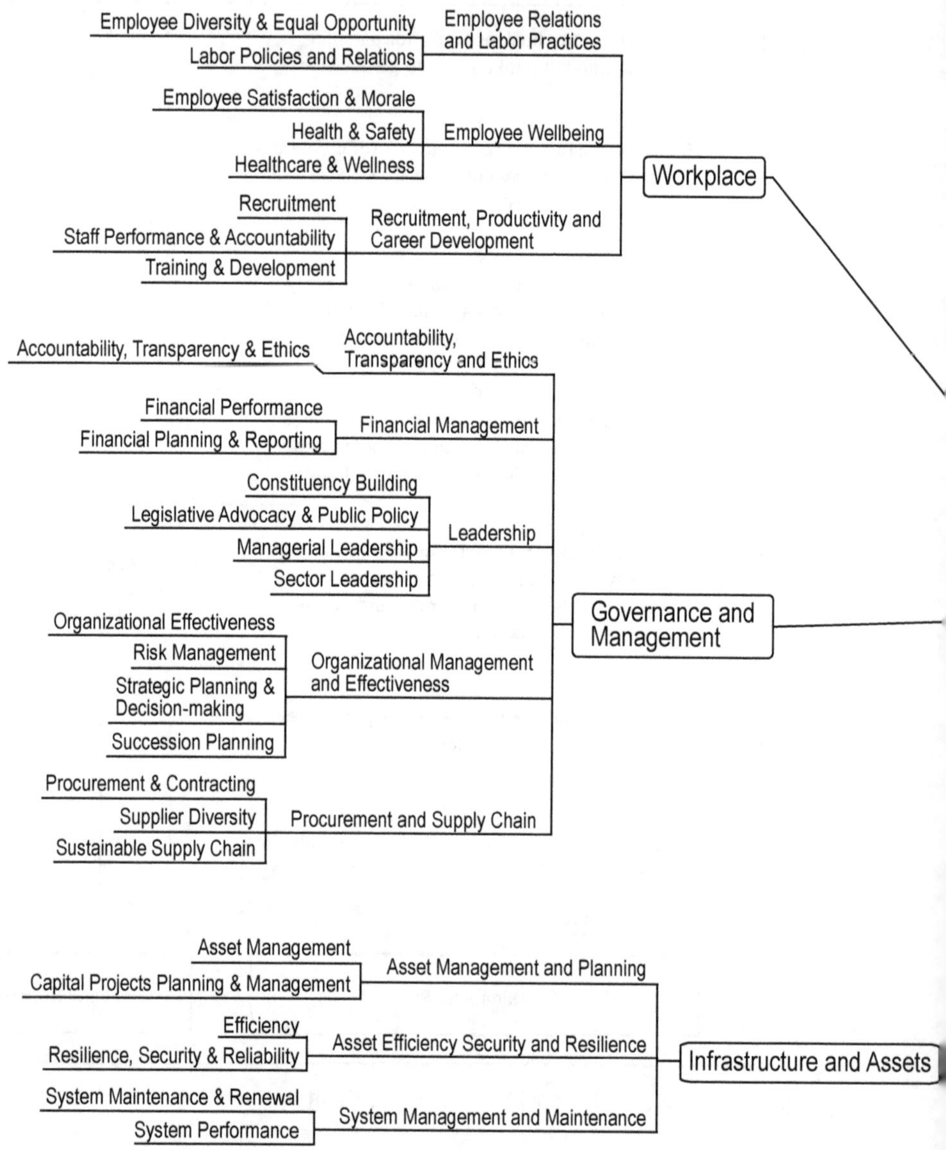

SFPUC MISSION STATEMENT:

To serve San Francisco and its Bay Area customers with reliable, high quality and affordable water wh maximizing benefits from power operations and responsibly managing the resources entrusted to i care

To protect public health and the aquatic environment by providing safe, reliable and efficient collectic treatment, and disposal of San Francisco's wastewater

To conduct its business affairs in a manner that promotes efficiency, minimizes waste and assures rate-payer confidence

To promote diversity and the health, safety and professional development of its employees

Figure 3.2. San Francisco Public Utility Commission's sustainability plan framework

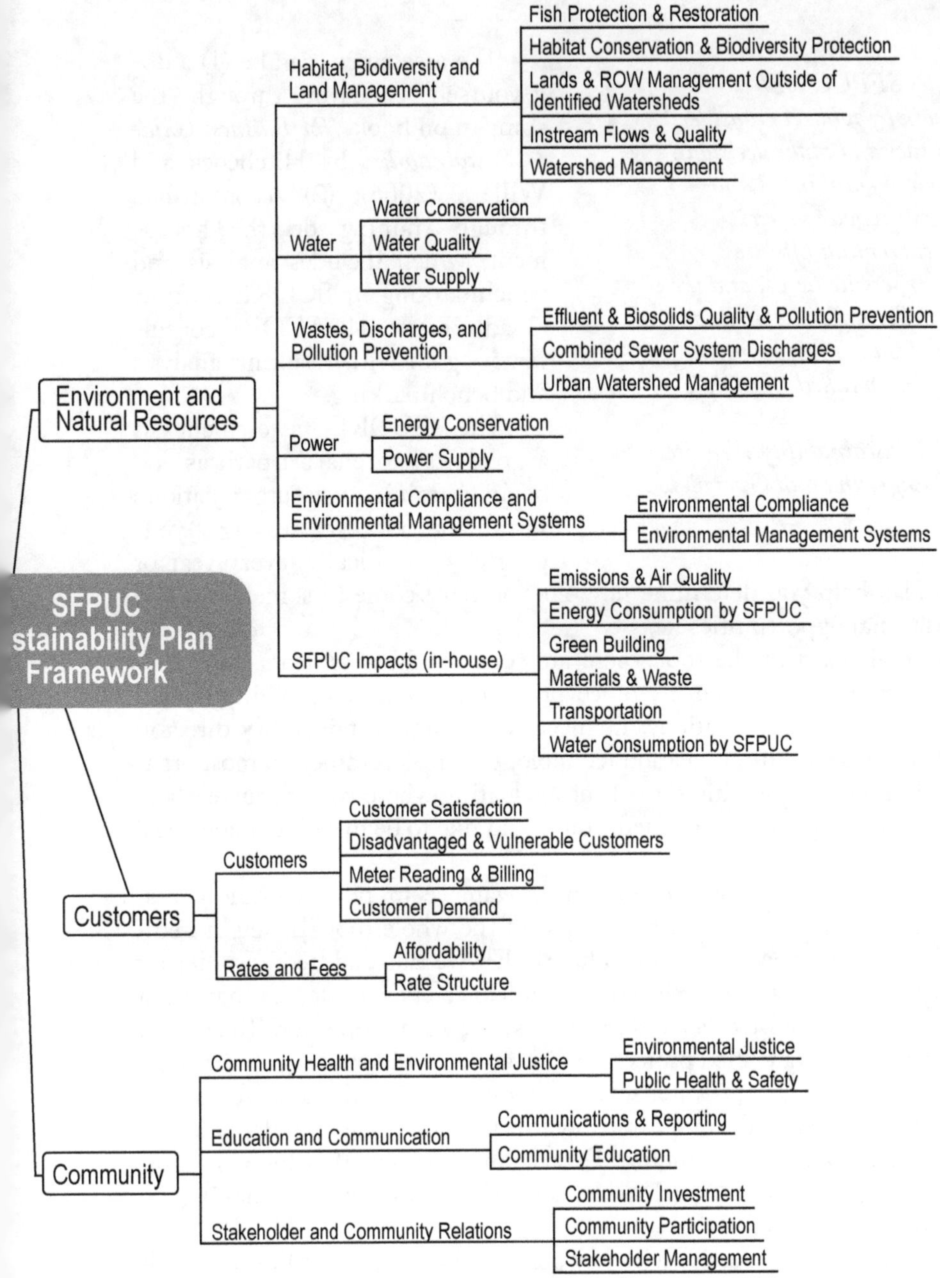

ORKING DEFINITION OF SUSTAINABILITY FOR THE SFPUC:

e framework through which SFPUC will responsibly manage the resources under its care, protect public health and balance its social and environmental responsibilities to the citizens and community, while providing cost effective services to its ratepayers.

[This] tool provides SFPUC with a strategic framework by which to consider cross-boundary issues and make decisions. It frames our departmental issues and decisions around our service sources and systems, our current and future customers and their service needs, and the integration of the effects of those needs on our long-term environment, economy, and communities (often called "the triple bottom line").

SFPUC Sustainability Web site, Sustainability Background and Overview

The assessment can be (1) a do-it-yourself endeavor through the companion book, *The Business Guide to Sustainability* by Hitchcock and Willard (2006); (2) accomplished through training on the assessment, which includes analysis and benchmarking by SCORE staff; or (3) administered by SCORE consultants, again with resulting analysis and benchmarking.

The SCORE assessment is a good way to chart a path as you begin examining your organization's sustainability. Reassessing your progress periodically (every year or two) may also help you determine how far you have come in some areas and where additional opportunities lie.

In the main part of the tool, there are functional assessments for executives, facilities, human resources, office practices, purchasing, environmental affairs, marketing/public relations, finance/accounting, sustainability director, and information and communication technology—areas common to most organizations. It is worth spending some time evaluating where you are currently in various areas and then deciding where you would like to be in a sustainably ideal world.

Appendix C lists only the questions in the water sector portion of the assessment, but it gives you an idea of the format of the whole tool. The eight items listed here are issues specific to our industry that we must address in order for our society as a whole to become more sustainable. They are topics that water utilities, more than most other organizations, are responsible for. They cover water conservation, watershed protection, life-cycle considerations, and more.

For each practice, in the water sector section as well as the basic functional assessments, you are asked to rate your organization on a 1 to 9 scale, based on examples given along the continuum. Few organizations qualify for 9s in most areas, but these provide our "North Stars," something to strive for in making our utilities more sustainable.

Going through this exercise with key personnel from your utility will likely spark some great conversations and hopefully contribute to a shift in thinking about what is possible in the water industry and in your utility in particular.

Backcasting

To discover what you envision a sustainable water utility to be, one great tool is something called *backcasting*. This process is heavily utilized by The Natural Step (TNS).

Sustainability Planning Framework/ Manual for Sydney, Australia

In March 2008, Sydney Water endorsed the Sustainability Planning Manual. The manual sits within Sydney Water's Environment Management System.

The manual outlines a decision-making framework to guide the user through any strategic planning process. Key features of the manual include:

- setting objectives, criteria, and measures at the outset;
- establishing a multidisciplinary project team;
- engaging with stakeholders and involving them in the decision-making process; and
- using a multicriteria analysis (MCA) as a robust and transparent way of ranking options.

The decision-making process outlined in Sydney Water's Sustainability Planning Manual is a six-step process based on the Water Services Association of Australia's (WSAA) Sustainability Framework: Methodology for evaluating the overall sustainability of urban water systems (WSAA Occasional Paper No. 17, 2008). Using this process to investigate options for projects under consideration allows Sydney Water to consider and rank a range of servicing options according to their environmental, social, economic, and technical elements, in consultation with key stakeholders. The MCA assists the project team in comparing options according to a set of objectives and criteria that have been agreed to by stakeholders as part of the consultation process. This helps to determine the most sustainable option and assists the project team through its ease of use, flexibility, and transparency.

The manual also contains information and provides links to additional Sydney Water resources, tools, and guidelines that the user may find helpful, such as information on stakeholder engagement, environmental risk assessment, and human health risk assessment.

The concept of backcasting (Figure 3.3) is central to a strategic approach for sustainable development. It is a way of planning in which a successful outcome is imagined in the future, followed by the question: "what do we need to do today to reach that successful outcome?"

Strategic sustainable development relies on "backcasting from sustainability principles." These principles are based in science and represent something we can all agree on, which is that if these principles are violated, then our global society is unsustainable.

The Natural Step system conditions for a sustainable society, as described in chapter 1, are described here:

"In a sustainable society, nature is not subject to systematically increasing:

- concentrations of substances extracted from the Earth's crust;
- concentrations of substances produced by society; or
- degradation by physical means.

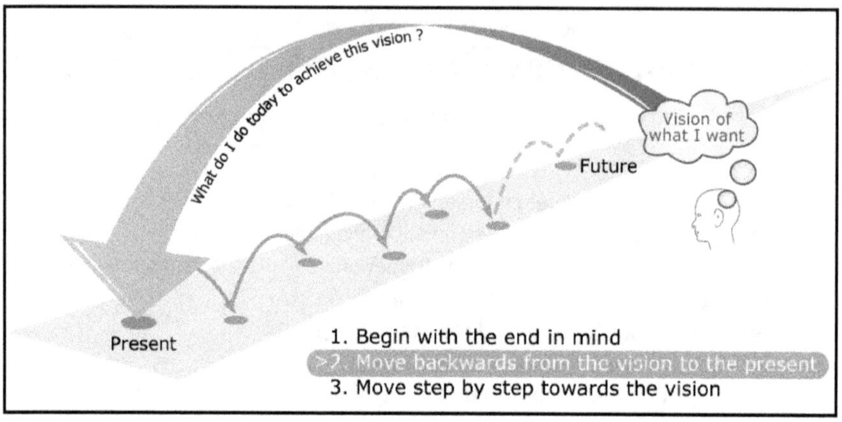

Figure 3.3. Backcasting is an important component of sustainable development

And, in that society. . .

- people are not subject to conditions that systematically undermine their capacity to meet their needs."

To achieve a sustainable society, we know we can't violate those principles. We don't know exactly what that society will look like, but we can define success on a principle level. Backcasting from principles is like playing chess—we don't know exactly what success will look like, but we know the principles of checkmate—and we go about playing the game in strategic ways, always keeping that vision of future success in mind (The Natural Step, 2009).

Process Flowchart

Most water utilities will have a basic flowchart similar to the one shown in Figure 3.4. We all provide water to our customers and contain the support structure to make sure that happens. This chart may not exactly mirror your structure, so adapt this to fit your organization or come up with something that feels more accurate for you. This type of chart will give you an idea of what needs to be tracked at your utility.

Major areas a water utility would likely assess include carbon emissions, materials usage, waste reduction, and sustainability progress. Tools to assess your current situation and track changes in these areas are outlined below.

Management Systems

To set priorities, plan projects, monitor progress, and evaluate the success of your sustainability efforts, you might employ what is often referred to as a *management system*. You may choose to adopt a formal environmental management system (EMS) that is ISO (International Organization for Standardization) 14001 conformant, adopt a broader sustainability management system (SMS), or you may choose to be more informal about your sustainability initiatives. The advantage of a management system is that it keeps things moving. Without a system

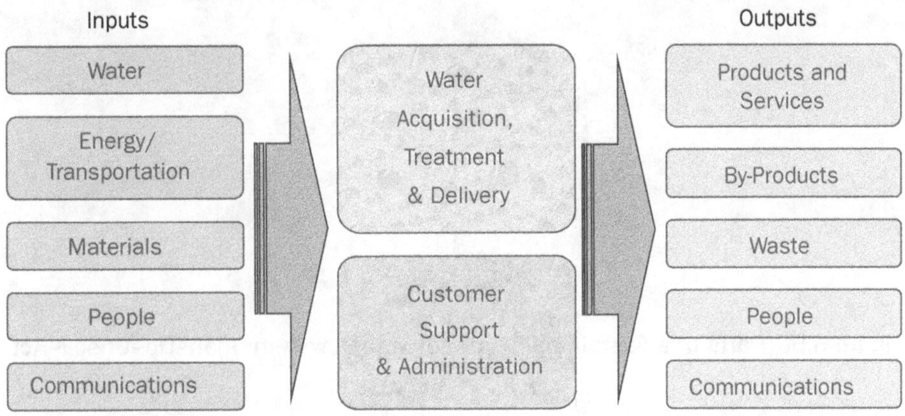

Figure 3.4. Example process flow diagram

in place, it is possible for sustainability initiatives to stall. If you don't adopt a management system, be sure you have a sustainability coordinator or champion of some sort to keep your utility progressing.

Environmental management system. An EMS is a set of processes and practices that enable an organization to systematically assess and manage its environmental "footprint," i.e., the environmental impact associated with its activities, products, and services. It is an important vehicle for improving environmental performance by providing organizations with the tools to successfully manage their environmental activities in a cost-effective manner. EMSs come in many forms and can be used in a variety of sectors. Most EMSs are built on the "Plan, Do, Check, Act" model (see Figure 3.5) (USEPA, 2009).

ISO has a "14001" set of requirements for an EMS. Fulfilling these requirements requires objective evidence that can be audited to demonstrate that the environmental management system is operating effectively in conformity with the ISO standard. These standards primarily focus on improving your organization's environmental performance and regulatory compliance.

The standards can also be used to meet external objectives, such as providing assurance on environmental issues to external stakeholders (e.g., customers, the community, and regulatory agencies) and supporting your claims and communication about your environmental policies, plans, and actions (ISO, 2009).

Evaluating your environmental impacts for an environmental management system can be complex and time-consuming but can be valuable if the process is followed precisely. Should you decide to get ISO 14001 certification, go to www.iso.org for specific instructions on how to go through the process.

Sustainability management system. To go beyond environmental considerations, consider adopting an SMS. It takes a more holistic view, including economic, environmental, and social elements, looking less at "how can we do what we do better," and more at "what do we need to do differently."

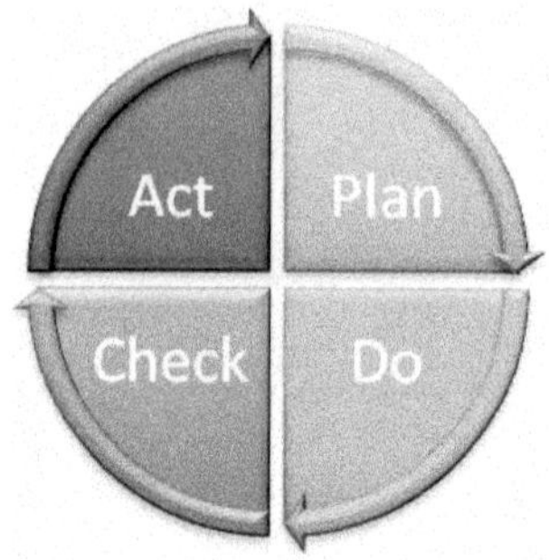

Figure 3.5. Parts of a sustainability management system: Plan–Do–Check–Act

Like an EMS, an SMS usually has four parts:

1. Plan: Identify and select improvement projects. Create a process or structure for how to continue the planning process into the future.
2. Do/Implement: Launch the projects and develop appropriate support systems, such as training, defined roles and responsibilities, communication, and document management.
3. Check/Monitor: Check progress of the effort, monitor support systems, and take preventive and corrective action as needed.
4. Act/Review: Evaluate the effectiveness of the projects and the initiative overall in a top management strategic review (Hitchcock, 2002).

The benefits of an SMS include the following:

- moves from single sustainability initiatives to an integrated business approach,
- provides a systemic framework for moving towards sustainability,
- becomes more proactive rather than reactive in dealing with environmental affairs,
- transfers learning from project to project,
- improves environmental performance,
- uncovers cost-saving ideas,
- improves environmental compliance with regulations,
- reduces risks and liabilities, and
- exceeds customer expectations (Hitchcock, 2002).

Resource Usage

Process Mapping

From an environmental perspective, sustainability is primarily about inputs and outputs. To systematically evaluate your utility's use of resources, you might start with a process mapping exercise. Ideally, this requires the participation of all employees, so it can be a big commitment. The sidebar on p. 33 details Tualatin

Process Mapping at TVWD

Five District staff went through a series of The Natural Step workshops in 2004 to learn how to implement process mapping and backcasting at the District. In January 2005, the workshop participants began to put these processes into action. All 100+ TVWD employees attended a half-day session to gain a background in sustainability and The Natural Step.

The staff was then divided into 19 work groups. Groups were chosen based on current crews and program staff—construction crew, accounting department, meter readers, etc. Some groups met three or four times in order to complete the process, others were done in one or two. It varied depending on the number of different types of tasks in each group.

The five trained workshop participants first met with each group to identify what the group does, what resources it consumes, and what waste it produces. Each group first listed what activities were undertaken every day by each employee, summarizing tasks when possible. They then broke each part of the process into substeps to make sure they knew exactly what went into each step. (See Figure 3.6, center column.)

Each group inventoried what physical materials (inputs) were used in each process, identifying over 400 items amongst all groups. (See Figure 3.6, left column, for example of inputs.) They also noted the waste or output from each part of the process. This set of meetings revealed the frequency of usage, cost, and degree of influence the District has over the purchase of these items and areas staff thought could be investigated for improved sustainability.

Though time-consuming, simply discussing these steps instantly yielded employee suggestions for ways to reduce the use of resources, increase operational efficiency, and reduce costs. Many of those ideas were implemented, processes modified, and waste reduced or eliminated on the spot.

It is rare to have the opportunity to meet with every employee in an organization, and the workshop participants took full advantage of the access to employee input and suggestions, soliciting information regarding working relationships and efficiencies as well as resource data. The information collected in these meetings was synthesized into a strategic plan that was reviewed by the TVWD Sustainability Steering Committee and approved by the District's five-member elected Board.

Each of the identified inputs was rated using The Natural Step system conditions and assigned a ranking indicating how important it was for the District to examine. This part of the process took into account such attributes as a material's toxicity, whether or not it was extracted from the earth, its cost, the number of alternatives available, and other considerations. (See Figure 3.7.)

First, this chart gave us an idea of what items might yield the most return for our time in investigating alternatives. Second, in completing this exercise, we included all toxic or hazardous substances used in our operations. We filtered the spreadsheet for those substances and created a new database of all the ingredients in each hazardous substance. This can be done manually and entered into something like Microsoft Access® or fed to any of several companies that focus on environmental health and safety, provide software to organize material safety data sheets (MSDS), and highlight products with specific ingredients you'd like to remove from your shelves.

(continued p. 35)

Inputs	Process: Meter Reading Substeps	Outputs
computer cradle/charger/unit paper (Green Bar) printer electricity phone	**Load Routes** Receive phone request Printed copy of request	filed by Pete for 6 months then wastepaper
paper printer	**Print out cycles** Print copy of map, if needed Sort requests into efficient order	wastepaper (2 Green Bar copies) filed, then waste paper
vehicle fuel personal protective equipment batteries paint wood stakes	**Drive** Drive to site Use locator to find service line Paint markings	batteries vehicle emissions tires paint cans, recycled
phone vehicle fuel	**Read meter** Talk to contractor if issues arise (might just show contractor location if raining and paint won't stay)	
paint paper door hangers	**Door hangers, marking meter lids/curbs** File requests sheets (with notes) for three months, then recycle paint markings	wastepaper paint cans, recycled paint emissions (volatile organic carbons)
	Vehicle inspection and cleaning File requests sheets (with notes) for three months, then recycle	

Sample of one process divided into substeps, with inputs and outputs identified.
Sustainable/efficient procedures of note already in place:
- Double up in vehicles when possible
- Working on electronic archiving of reports, plans, maps

New sustainable/efficient opportunities:
- Use 8.5×11 instead of Green Bar—use less paper
- Notify people in other ways besides door hangers

Figure 3.6. Looking at a single job using process mapping—meter reading

We got the most benefit from the process itself, i.e., discussions about whether or not a particular paper report was needed or how to make a specific process more efficient. The next most helpful part for our organization was the hazardous substances information. We were able to make a comprehensive list, searchable by ingredient, that showed us, first and foremost, that we had very few hazardous substances to worry about. But those we do have now come to the top of the list and we can be on the lookout for alternatives. Unfortunately, the top substances at present are gas and oil, for which we have no good alternatives.

The actual chart in Figure 3.7 has been less useful to us. We have used it to consolidate some items that were being purchased in several different brands. And we developed a process whereby staff wishing to use new products fills out a short form that is submitted to our safety coordinator. He then is given an opportunity to see if we have other substances that would be appropriate for the use or if he knows of any other safer, more environmentally friendly alternatives.

In this instance, the product (the spreadsheet) was helpful, but not as valuable as the process itself and the changes and ideas that it sparked.

Valley Water District's experience completing this process, with sample tables and lessons learned.

If you are not able to go through such an exercise, it is still valuable to track the usage of your major resources, which are likely energy, water, and labor.

Energy Usage

Most water utilities are pretty energy-intensive. Treating water; pumping it to customers at higher elevations; pumping it into and out of aquifer storage and recovery (ASR) wells; and heating, cooling, and lighting our facilities are all major expenses and uses of this resource.

The relationship between energy use and water delivery are mutually reinforcing, and utilities benefit from closely examining energy use internally and energy use driven by water demand. Steve Conrad, a water-energy nexus researcher at Simon Fraser University, British Columbia, has observed significant opportunity for utilities to improve their management of energy and has contributed information on energy issues in utilities (see p. 41). A more in-depth review can be found in appendix D, which provides plenty of further reading, should you want to delve deeper into this topic.

Appendix D identifies literature related to management, improvements, and planning for future energy and greenhouse gas performance in the following areas:

- pumping,
- pressure reduction and leakage,
- new water technologies,
- new wastewater technologies,
- decentralized systems,

		"The Natural Step" System Conditions											
Work Group	Input	Extraction	Persistency	Toxicity	Biodiversity	Efficiency	Equity	Cost	Subtotal	Frequency of Activity	Degree of Influence	Multiplied	RANKING
Construction	blades	5	4	2	4	4	4	3	26	5	5	650	1
Fleet	filters	5	5	3	5	3	3	1	25	5	5	625	2
Safety	safety glasses	5	5	5	5	4	5	2	31	5	4	620	3
Valve	markers	5	5	5	5	4	5	1	30	4	5	600	4
Engineering	pens	5	5	5	5	4	5	1	30	4	5	600	4
Fleet	tires	5	5	5	5	5	3	2	30	4	5	600	4
Building and grounds	blades (plastic)	5	5	4	5	5	3	2	29	4	5	580	7
Construction	batteries	5	5	4	5	5	3	1	28	4	5	560	8
Engineering	CDs	5	5	4	5	5	3	1	28	4	5	560	8
Customer Service	air fresheners	3	3	4	4	4	3	1	22	5	5	550	10
Building and grounds	garbage bags	5	5	5	5	4	5	2	31	4	4	496	11
Fleet	lubricants	5	5	5	5	5	3	3	31	4	4	496	11
Valve	PVC	5	5	5	5	4	5	2	31	4	4	496	11
Fleet	transmission fluid	5	5	5	5	5	3	3	31	4	4	496	11
Safety Emergency	raingear	5	5	5	5	4	5	4	33	3	5	495	15
Safety Emergency	gloves-latex	5	5	5	5	4	5	1	30	4	4	480	16

Figure 3.7. Looking at utility-wide resources usage using process mapping

Energy is required for all facets of the urban water cycle, from the collection, pumping, and treatment of supply water, to transmission and distribution, to wastewater pumping and treatment (AwwaRF, 2007).

Energy decisions made by utilities may include capital plans, forecasting future water supplies, improvements in regional operations, and investments in alternative energy and water and wastewater technologies. Water and wastewater sectors actually make a relatively small contribution to greenhouse gas emissions compared with other sectors. However, individual utilities can be some of the largest users of energy at a company level, and energy-intensive sources of water may increase energy use in the future.

In the United States, 3% of total electricity generated, or 75 billion kW·h per year, is used by publicly owned water and wastewater utilities (Cohen, Nelson, and Wolff, 2004). It is estimated that energy use in the water and wastewater sector will grow by 40% over the next 20 years, which will be driven by new regulations and population growth (Arpke and Hutzler, 2006).

In Australia, the water and wastewater sector produces approximately 0.5% of greenhouse gas emissions, excluding diffuse emissions (Foley and Lant, 2007). A number of scenarios of future water supply from energy-intensive sources indicate that energy use for utilities may increase significantly but will still not become a large contributor in the context of the Australian economy (Kenway et al., 2008).

Urban water systems will rely on substantial and increasing amounts of energy. Many management strategies focus on the energy of water and wastewater treatment and pumping but omit water end use. For instance, total emissions through the full life cycle of the water "product" are often overlooked (Racoviceanu et al., 2007; Kenway et al., 2008).

This literature identified various steps utilities are taking to reduce energy and GHG emissions that include (WSAA, 2007):

- avoiding energy use where other options exist to achieve the same outcome,
- improving energy efficiency measures (e.g., deploying advanced control for operations optimization, implementing energy management systems, and installing variable speed pumps rather than fixed speed),
- using wastes to generate renewable power, and
- sequestering carbon and purchasing carbon offsets.

Utilities are beginning to develop life-cycle inventories focusing on energy use and GHG emissions. The City of Toronto, for instance, developed inventories for three processes within its water treatment plants. Inventories were based on chemical production, transportation of materials, and water treatment plant operations (Racoviceanu et al., 2007). Following the initial assessment, operations accounted for 94% of total energy use and 90% of GHG emissions.

- relation between energy pricing and water consumption,
- water use for energy supply,
- demand management and efficient end use,
- energy recovery,
- influence from climate change, and
- operational energy decisions.

Energy efficiency is probably the cheapest, easiest renewable resource we have. It is a fairly easy place to start tracking your resource usage. If you have few sites, just collect your electricity bills and use Excel® to create a spreadsheet of various types of uses. If you have many facilities, you may be able to get a consolidated statement from your electric utility company to simplify this process.

Divide your usage into water treatment facilities, pump stations, administrative buildings, reservoirs, and other facilities, such as fluoride stations and vaults. Make sure your biggest electricity uses are called out so you can explore opportunities to reduce that usage. When you have multiple years of data, you can also see how usage at each type of facility has changed over the years (Figure 3.8).

Your electric utility or a nonprofit organization in your area might also conduct an energy audit to help you identify opportunities at your particular facilities.

See case studies on pp. 41 and 44 and chapter 4 for a variety of actions that can reduce your energy usage.

Water Usage

It is critical for our organizations to promote water conservation among our customers and to practice it ourselves—it is obviously one of the water industry's most important resources. In particular, reducing peak usage is important, as it allows us to reduce the size of our infrastructure. We can help ourselves by making sure our operations do not add to summer peaking. Miami-Dade Water and Sewer District's Five Year Plan (Figure 3.9) lists practices in which most water utilities are engaged and whether they affect the average-day or peak demand.

As noted earlier, AWWA and others have produced many excellent materials that detail water conservation, and you will find some of those referenced at the end of this chapter. For that reason, this section is very brief. I encourage you to explore those materials if you need more information regarding the value of water, how to communicate that value, and ways to use water most efficiently. It is critical that we set a stellar example in this area.

From a sustainability standpoint, however, it is most important for us to consider (1) our utility's water efficiency in providing quality water to our customers and (2) whether or not there are other models that will allow us to use water differently than we are used to. How can we minimize the amount we move water around and maximize the use we get out of the water we collect? Throughout this book you will find examples of water utilities that have found innovative answers to these questions.

Another thing to consider is the concept of a water footprint. An organization's water footprint is the total volume of fresh water used to produce the goods and services consumed by that organization. There is not as much information on this topic as there is on carbon footprints, especially for businesses, but it is one more way to calculate the resources we consume and take into account the water used to create the items we use in our operations. Clear standards regarding exactly what should go into a water footprint do not yet exist. However, two Web sites are listed in the References section that will give you more information about water footprinting.

Electricity Usage and Cost		2007			2008		
Facility	**Address**	**KWh Use**	**Cost**	**Percent**	**KWh Use**	**Cost**	**Percent**
Pump Stations							
126th Pump Station	5474 NW 126th Terrace	4,592	$557	0%	2,228	$342	0%
189th Reservoir / Pump Station / Well	7400 SW 189th Avenue	555,500	$49,201	**14%**	572,700	$47,068	**14%**
Bethany Pump Station	14517 NW Laidlaw Road	304,160	$29,830	8%	313,520	$28,060	8%
Bonny Slope I Reservoir / Pump Station	10465 NW Laidlaw Road	10,301	$1,073	0%	10,507	$1,079	0%
Catlin Crest Pump Station	7340 SW Barnes Road	38,775	$3,658	1%	43,821	$4,090	1%
Cooper Mountain Reservoir / Pump Station	18888 SW Kemmer Road	273,274	$21,718	7%	355,776	$26,843	9%
Florence Lane Reservoir / Pump Station	6700 SW Florence Lane	21,678	$2,142	1%	27,871	$2,608	1%
Goyak Reservoir / Pump Station	18700 SW Rigert Road	553,700	$52,841	**14%**	596,200	$50,941	**15%**
Inglewood Reservoir / Pump Station	9500 SW Inglewood Street	383,300	$38,350	**10%**	346,600	$33,225	9%
Power Generator / Pump Station	11285 SW Center Street	17,955	$1,806	0%	18,364	$1,823	0%
Power Generator / Pump Station	11285 SW Center Street	254	$34	0%	96	$201	0%
Somerset Reservoir/ Pump Station	5425 W Burnside Street	1,128	$250	0%	1,337	$263	0%
Sunset Reservoir / Pump Station	8185 Sunset Highway	833,200	$73,302	**21%**	832,100	$72,185	**21%**
Tueffel Reservoir / Pump Station	7323 SW Barnes Road	151,512	$13,767	4%	20,560	$1,959	1%
Viewmont Pump Station	601 SW Viewmont Drive	16,146	$1,639	0%	15,796	$1,596	0%
	Subtotal:	**3,165,475**	**$290,168**	**82%**	**3,157,476**	**$272,283**	**80%**
Administration Offices							
TVWD Main Office	1850 SW 170th Avenue	600,720	$48,489	15%	605,280	$48,998	15%
Metzger Office	6501 SW Taylor's Ferry Road	18,893	$1,883	0%	19,966	$1,966	0%
Metzger Office	6501 SW Taylor's Ferry Road	1,592	$289	0%	1,584	$255	0%
Metzger Office	6501 SW Taylor's Ferry Road	3,538	$460	0%	3,528	$453	0%
	Subtotal:	**624,743**	**$51,121**	**16%**	**630,358**	**$51,672**	**16%**
Reservoirs							
Bonny Slope Park Reservoir	11140 NW Thompson Road	442	$187		676	$204	
Cornell Reservoir	8261 NW Cornell Road	8,530	$904		7,464	$804	
Garden Home Reservoir	6207 SW Garden Home Road	799	$156		185	$160	
Garden Home Reservoir	6207 SW Garden Home Road	1,255	$261		792	$139	
Grabhorn Reservoir / Well	7975 SW Grabhorn Road	7,869	$806		107,920	$17,382	
Hyde Park Reservoir	10001 SW Ardenwood Street	1,284	$262		879	$222	
North Road Reservoir	4280 North Road	886	$227		1,126	$244	
North Road Reservoir	4240 North Road	55	$152		43	$148	
Rosander Reservoir	8560 SW Barnes Road	3,412	$450		628	$200	
Schell Reservoir	9755 SW Ruesser Road	36	$149		55	$149	
Springville Reservoir	14160 NW Springville Road	2,869	$400		1,451	$274	
Thompson Reservoir	12520 NW Thompson Road	156	$161		49	$148	
	Subtotal:	**27,593**	**$4,115**	**1%**	**121,268**	**$20,074**	**3%**
Other Facilities: Meters, Flow Valves, Fluoride Stations							
221st JWC inter-tie / Fluoride Station	2787 SW 221st Avenue	5,972	$675		9,321	$968	
Cornelius Pass Regulator / Fluoride Station	4375 SW Cornelius Pass Road	21,440	$2,196		22,560	$2,776	
Garden Home Meter Vault	65th Avenue & Garden Home Ro	237	$168		263	$167	
Pressure Regulation Valve Vault	Inglis Drive	355	$177		73	$139	
Emergency Tie-in with Portland Water Bure	6210 SW Multnomah Boulevard	583	$201		619	$199	
Beaverton Inter-tie	12159 SW Center St	334	$176		317	$172	
TVWD Main Office North Gate	170th Avenue & Merlo Road	325	$175		320	$172	
Beaverton Tie-in	153rd & Jenkins Road	136	$158		131	$156	
Pressure Regulation Valve Vault	11375 SW Center Street	921	$229		7,328	$792	
Metzger Meter Vault	SW 80th & Florence	6,701	$746		6,491	$722	
Old Metzger Meter Vault	4824 SW Olsen Road	0	$147		0	$144	
Pressure Regulation Valve Vault	11864 NW Blackhawk	1	$147		0	$144	
Pressure Regulation Valve Vault	Walker Road & Meadow	162	$161		32	$147	
Portland 54" Flow Meter / Valve	7412 SW Beaverton Hillsdale Hig	3,127	$425		2,644	$379	
Radio Tower	8935 SW Mayberry Place	3,697	$476		3,443	$450	
Shuepbach Well	5640 SW 160th Avenue	0	$198		322	$221	
	Subtotal:	**43,991**	**$6,455**	**1%**	**53,864**	**$7,748**	**1%**
	Total:	**3,861,802**	**$351,859**	**100%**	**3,962,966**	**$351,777**	**100%**

		2007 KWh Use	2007 Cost	2007 Percent	2008 KWh Use	2008 Cost	2008 Percent
	Pump Stations	3,165,475	$290,168	82%	3,157,476	$272,283	80%
	Administration Offices	624,743	$51,121	16%	630,358	$51,672	16%
	Reservoirs	27,593	$4,115	1%	121,268	$20,074	3%
	Other Facilities	43,991	$6,455	1%	53,864	$7,748	1%
	Total:	**3,861,802**	**$351,859**	**100%**	**3,962,966**	**$351,777**	**100%**

Figure 3.8. Electricity use by type of facility

Measures Included in Five-Year Plan	
Conserve Florida Measures	Effect on Demand*
Source water metering	B
Service-connection metering	B
Fixed-interval meter reading	B
Conservation rate structures	B
Informative billing	B
Test, calibrate, repair, and/or replace meters	B
System audit	A
Leak detection and repair	A
Selective usage of pressure-reducing valves	A
Conservation coordinator	B
Wholesale water supplier assistance programs	B
Landscape/irrigation ordinance development and implementation	P
Public information/education	B
Water bill inserts	B
Conserve Florida Quantifiable Best Management Plans	
Water-efficient landscape and irrigation evaluations and rebates	B
High-efficiency clothes washer rebates	B
Ultra-low-flush toilet rebates	B
Retrofit kit give-away	A
Industrial, commercial, and institutional water use evaluations/ implementation	B

A = affects average-day demand
P = affects peak demand
B = affects both average-day demand and peak demand

Figure 3.9. Miami-Dade Water and Sewer District's conservation measures

Human Resources Usage

Hiring well, training well, and planning for the future is a sustainable way to manage the human resources essential to a water utility.

Many current job applicants are looking for employment at organizations they believe will respect them and provide a healthy working environment.

Based on a 2009 survey of water and wastewater agencies and utilities in six Bay Area (California) counties, employers there are projected to need as many as 677 new and replacement workers in seven mission-critical occupations over the next five years.
BAYWORK/Centers of Excellence

Include your focus on sustainability in job announcements to attract solid job

Metro Vancouver, and its affiliate, the Greater Vancouver Water District (GVWD), is a Canadian water utility serving the 2 million people in its 22 member municipalities in the greater Vancouver, British Columbia, area. Metro Vancouver utilizes resources in the mountainous Capilano, Seymour, and Coquitlam watersheds. Water is delivered through 550 kilometers of water mains, 22 peaking reservoirs, and 16 booster pump stations, and GVWD also collects and treats the region's wastewater streams.

GVWD owns and operates one water treatment plant, designed to treat 1.8 billion liters (475 million gallons) per day, and five wastewater treatment plants. The majority of GVWD operations are water treatment and transmission and wastewater treatment. Member municipalities are responsible for direct consumer water distribution and wastewater collection.

As an organization, Metro Vancouver has committed to sustainability as an underlying business principal. It is the core vision for the organization, with the ultimate goal for Metro Vancouver to be the greenest region in the world by 2020. The Sustainable Region Initiative (SRI) identifies the need for a series of management plans to address delivery of services according to the principles of sustainability. The SRI has three sets of operating principles:

- protect and enhance the natural environment,
- provide for ongoing prosperity, and
- build community capacity and social cohesion.

To help define and implement the SRI, Metro Vancouver's Drinking Water Management Plan (DWMP) was adopted in 2005. The goals of the plan are to:

- provide clean, safe drinking water;
- ensure the sustainable use of water;
- ensure the efficient supply of water; and
- manage and protect the watersheds that provide the region's water as natural assets.

The region is moving toward managing water by encompassing the full water cycle. This means considering the relationships between stormwater, drinking water, and liquid waste that have traditionally been treated as separate water management issues. The DWMP includes actions to match water quality usage requirements by assessing alternative sources of water (rainwater, grey water, and wastewater) for nonpotable use, eliminating once-through cooling water, requiring water-efficient fixtures in new construction and renovations, implementing leak identification and repair programs, enforcing the Water Shortage Response Plan, developing residential water metering programs, establishing municipal rebate programs for water-efficient fixtures and appliances, and assessing the merits of standardized industrial, commercial, and institutional water audits for the largest 25% of business users to initiate water conservation improvements.

To further reduce resources and energy needed to operate the water treatment and distribution system, energy management is a fundamental decision to meet Metro Vancouver's sustainability framework targets and priorities, including:

- energy: be a net contributor of energy by 2015;
- greenhouse gases: be carbon neutral by 2012, excluding solid waste operations;
- greenhouse gases: reduce regional greenhouse gases by 15% by 2015 and 33% by 2020; and
- waste: increase energy from liquid waste by 10% by 2012.

(continued)

To help meet these goals and reduce energy consumption at Metro Vancouver's water system, the GVWD partnered with the Power Smart Division of BC Hydro. The Power Smart Partner Program provides eligible organizations with the opportunity to partner with BC Hydro and gain access to a variety of tools and resources to become more energy efficient.

Metro Vancouver assesses all energy projects for their implications on a range of sustainability indicators, including economic, environmental, social, and technical impacts (as in triple bottom line accounting). As part of this, all energy projects are financially assessed using life-cycle costs and benefits, and the related reporting indicators of benefit-cost ratio, net present value, and internal rate of return. Environmental indicators and assessments always include, but are not limited to, greenhouse gas emissions.

Metro Vancouver uses a variety of tools to achieve their energy management goals, including:

- hydraulic modeling to assess operations in terms of optimizing energy use;
- operations optimization advanced control to manage energy and water supply;
- sustainable purchasing and business case tools to help identify project selections that balance environmental, social, and economic benefits; and
- GHG accounting using protocols from IPCC, World Business Council for Sustainable Development/World Resources Institute, and BC government.

Metro Vancouver partnered with BC Hydro to evaluate the energy savings during off-peak times at two pump stations—Central Park and Cape Horn. These pump stations were built in the mid-1970s using fixed speed motors when energy use was not a consideration in their original design.

Metro Vancouver constructed a custom GHG calculator and economic analysis tool to assess the feasibility of pump replacement and system configurations to reduce energy use. Following initial analysis, system engineers determined that energy reduction was possible by using demands downstream of the Central Park pump station to cycle the reservoir, thus eliminating all pumping, and to use the low motor speed at the Cape Horn pumping station during the off-peak times to save electrical energy.

Potential barriers include being unable to submit projects for external funding at a pace that matches how analyses and approvals need to take place. In addition, Metro Vancouver believes that the low price of electricity locally acts as a disincentive to innovation for certain kinds of projects (such as distribution of captured heat from cogeneration).

Contributed by Steve Conrad, Simon Fraser University.

applicants. When you get prospective employees who are interested in investing in your utility, you get people you can feel good about investing in, as well.

Motivating and training utility workforces has a body of literature all its own and is not the focus of this book, except to say that keeping employees engaged in their work is essential. Once you have invested in initial training, keeping your current talent is the most sustainable path. Even when economic times are tight, it is probably more cost-effective to spend resources on keeping a current employee engaged than hiring a new one. An interesting seminar or new employee benefit might be helpful in that regard.

In October 2009, Kelly Services, a global workforce solutions organization, released the results of a survey that found employees across all age generations gravitate to organizations considered ethically and environmentally responsible, with baby boomers (aged 48–65) more discerning than their younger colleagues in Gen Y (aged 18–29) and Gen X (aged 30–47).

The Kelly Global Workforce Index obtained the views of nearly 100,000 people in 34 countries.

Kelly Services vice president and managing director of Canadian operations, Karin French, said, "Employees take pride not only in what they do while at work but in what their organization stands for and how it is perceived by the entire community. It provides a sense of fulfillment to be part of an enterprise that is focused not only on performance outcomes but also on larger social goals."

Among the key findings of the survey:

- 85% of respondents are more likely to want to work for a company that is considered ethically and socially responsible.
- 80% are more likely to want to work for a company that is considered environmentally responsible.
- 74% say that in deciding where to work, an organization's reputation for ethical conduct is "very important."
- 53% would be prepared to accept a lesser role or a lower salary to work for an organization with a strong environmental and community conscience.
- 36% say that in deciding where to work, policies aimed at addressing global warming are "very important" (Kelly Services, 2009).

At the end of employment, workforce planning is essential for utilities, now, as well. An increasing number of utility employees are retiring and taking years or decades of knowledge and understanding with them.

The solution to the changing workforce problem may be as simple as getting an employee's daily duties down on paper, or as complex as implementing a workforce planning program, computerized maintenance management system (CMMS), or asset management system to track physical assets and systematize maintenance activities. However you do it, finding ways to document processes and make knowledge available to future staff will boost the efficiency, and therefore sustainability, of your staff now and in the future. AWWA has archived several good articles regarding workforce planning at the Water Library portion of their Web site (http://apps.awwa.org/WaterLibrary/Search.aspx).

One of the trade-offs in developing alternative water supplies is that they invariably involve greater use of energy. Unless that energy comes from a sustainable source, it is likely contributing to climate change.

Douglas Yoder, Miami-Dade Water and Sewer Department

Sustainable solutions are often achieved through public–private partnerships. The City of Toronto, for example, partnered with private company Enwave to develop the largest cooling district in North America created from a renewable resource. Connecting over 300 buildings, Toronto's deep lake water cooling system is an example of resource sharing that both improves the operations of utility infrastructure systems and serves broader community sustainability goals.

Deep water cooling involves using naturally cold water as a heat sink in a heat exchange system, eliminating the need for conventional air conditioning. The cold water is drawn from near the bottom or below the thermocline of a nearby water body. Toronto's deep lake water cooling system began operation in July 2004 and helps keep downtown Toronto buildings cool year-round. Water is drawn from 5 km (3.1 mi) out in Lake Ontario and from a depth of 83 m (272 ft) at a temperature of only 4°C (39.2°F). (See Figure 3.10.)

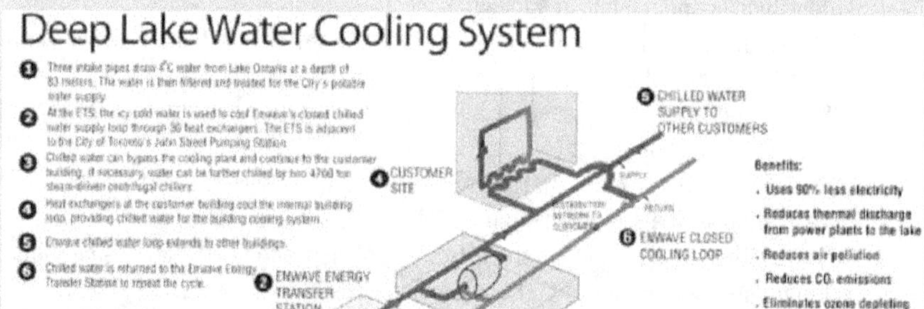

Figure 3.10. Enwave cooling system

The cold energy from the lake is used to cool water in a closed chilled water supply loop. The coldness of the lake water is transferred through 18 pairs of stainless-steel heat exchangers (with over 800 sandwiched plates per heat exchanger) to Enwave's closed-loop chilled water supply distribution network. Returning water from the distribution system is cooled using the deep chilled water from the lake and then by centrifugal polishing chillers to around 3°C (37.4°F). Once it has passed through the heat exchanger, at a temperature of 13°C (55.4°F), the water drawn from the lake then enters the city's potable water supply system. The system is designed to provide air conditioning of nearly 3.2 million m^2 (34.4 million ft^2) of office space, eliminating building electrically driven chillers. The systems can provide 75,000 tonnes of connected refrigeration demand.

The Enwave system reduces energy consumption by 90% and has additional benefits of reducing emissions of ozone-depleting substances and greenhouse gases. This cold energy of the subterranean loop has the capacity to eliminate harmful ozone-depleting refrigerants (CFCs and HCFCs) and remove 79,000 tons of carbon dioxide from the air—the equivalent of taking 15,800 cars off the road.

By building public confidence through strong internal leadership and an active communications program, the city was able to overcome potential barriers that often accompany large-scale capital investment projects. These include difficulties in allocating funding and convincing customers that third-party partnerships offer long-term stability and benefit to the city.

Contributed by Steve Conrad, Simon Fraser University

Carbon Emissions Inventory and Management Strategy Guidelines for Water Utilities

GHG Inventory

The other "end" of operational processes is waste and pollution. Chief among these are greenhouse gas (GHG) or carbon emissions. The Water Research Foundation recently published a report on GHG emissions (Huxley et al., 2009); the following information from that report details how to complete a GHG inventory and issues that are of particular interest to water utilities. An initial overview is included here, with more in-depth information in the case study on p. 48 and in appendix E.

> Minimizing climate change impacts is one of the key aspects of a comprehensive sustainability program, and documentation of your utility's greenhouse gas (GHG) emissions is fundamental to both development of effective strategies and measuring progress achieved. Preparing a GHG inventory and footprint analysis takes knowledge of local, state, and national regulations; GHG quantification protocols; as well as an understanding of your utility's philosophy about GHG reductions over the long haul.
>
> A full description of the process of preparing and updating a GHG inventory is beyond the scope of this text. However, the following synopsis of Water Research Foundation (Foundation) Project 4156, "Greenhouse Gas Emission Inventory and Management Strategy Guidelines for Water Utilities," will provide a snapshot of the process, as well as associated water-utility-specific considerations. This Foundation project resulted in a guidance document specifically tailored for water utilities, and it was intended to assist water utility staff in understanding GHG reporting programs, accounting principles, existing guidelines, and emission estimation methodologies for GHG source types relevant to water supply and treatment.

The GHG inventory process overview follows, and the concepts below are discussed in appendix E, but refer to the full WRF report for detailed descriptions of these topics.

- Understanding GHGs of interest
- Registries, programs, and protocols
- Accounting fundamentals
- Calculating emissions
- Water utility issues
- Management strategies

GHG Inventory Process Overview

There are many reasons that your water utility may begin to assess GHG emissions that are directly or indirectly caused by utility operations and to evaluate the projects or actions that may reduce or offset those emissions. These include, but are not limited to, existing and developing regulations requiring inventory and reporting of emissions, developing "carbon tax" and "cap-and-trade" regulation of GHG emissions, stakeholder initiatives demanding action, and environmental stewardship goals. In regards to regulatory impetus, the American Clean Energy and Security Act (ACES) was passed by the US House of Representatives in June 2009, and the companion Clean Energy Jobs and American Power Act is being developed in Senate subcommittee as of November 2009. Both bills propose a cap-and-trade system for GHG emissions and may impact some water utilities that generate GHG emissions over 25,000 metric tons/year. While your utility might not exceed this regulatory threshold, as was described previously in this book, there are a multitude of sustainability drivers that would create interest in establishing and maintaining a GHG emissions inventory.

Water utilities contribute to GHG emissions primarily through combustion of fuels in stationary sources or mobile sources (e.g., vehicles) and through consumption of electricity to perform the various functions necessary in the treatment and transport of water. Characteristics specific to your water utility, including source-water location, raw-water quality and the treatment required, size of the overall organization, and distribution system topography, will all affect the quantity of GHG emissions generated. By understanding the source of your GHG emissions and generating a baseline emissions inventory for your organization as a whole, your utility may be able to subsequently identify projects to reduce emissions.

However, under most circumstances, GHG emissions cannot be directly measured and thus translated into an inventory. You must instead calculate GHG emissions using measured "activity" data for various parameters (for example, fuel combusted, electricity consumed, or vehicle miles driven). These parameters are then typically multiplied by "emission factors" that estimate the quantity of GHGs emitted per unit of activity. As a utility, you will find that many protocols exist describing the preferred data sources, procedures for determination of emission factors, calculation procedures, and other issues required to produce accurate, complete, and transparent GHG emission inventories. You will also find that a fairly rigorous process is required for creation of "carbon offsets" from projects that permanently reduce your GHG inventory. However, to get from point A, or your intent to generate an inventory, to point Z, identification and implementation of ways to reduce your inventory, you will need to follow multiple steps, beginning with understanding what constitutes a GHG, and then by selecting an appropriate protocol to follow.

See appendix E for additional information.

WASTE PRODUCTION

Waste Assessment

In addition to emissions, there are other types of waste to measure. The most obvious is solid waste. One way to do a quick estimate of your solid waste issues is a waste assessment. This can vary in detail and accuracy, depending on your needs.

Some areas have extensive opportunities to recycle many materials. Other regions have limited or nonexistent recycling programs. Check with your waste hauler to see what can be recycled. Look also for nonprofit organizations that may have set up recycling depots not related to waste haulers. Even if you don't have a major recycling program in your area, it is still possible to do your own composting or take packing peanuts to a mailing center for reuse. This type of assessment may be just what you need to identify ways to reduce what goes from your operations to the landfill.

The easiest way to do a waste audit is to simply take a few bags out of the dumpster and see what is in them. Choose a quantity that you think will be representative of the facility you are auditing. Collect garbage that appears to be from as many different areas of the site as you would like to evaluate.

First, determine the total weight of the garbage you are about to evaluate. Then separate any items that can be recycled into categories, physically putting them into piles such as:

- corrugated cardboard,
- newspaper,
- mixed paper (light cardboard, copy paper, magazines, etc.),
- tin,
- aluminum,
- glass, and
- certain plastics (check with your recycler or waste hauler to see what they accept).

> TIP: Keep the results of your waste assessment nonpersonal. It doesn't matter who did this, just that the paper doesn't keep going in the trash.

United Utilities Planned Approach to Carbon Reduction Through 2050

Our ambition is to halve our greenhouse gas emissions by 2035, based on a 2005 baseline, and to reduce them by 60%, in line with government targets, by 2050, based on 1990 levels and as shown in Figure 3.11. While we have set ambitious targets, we have yet to identify how we will fully meet them. Figure 3.11 sets our carbon reduction plan broken down into four distinct phases, A–D. We will engage with our customers in each of these phases to ensure that they continue to support our carbon management approach.

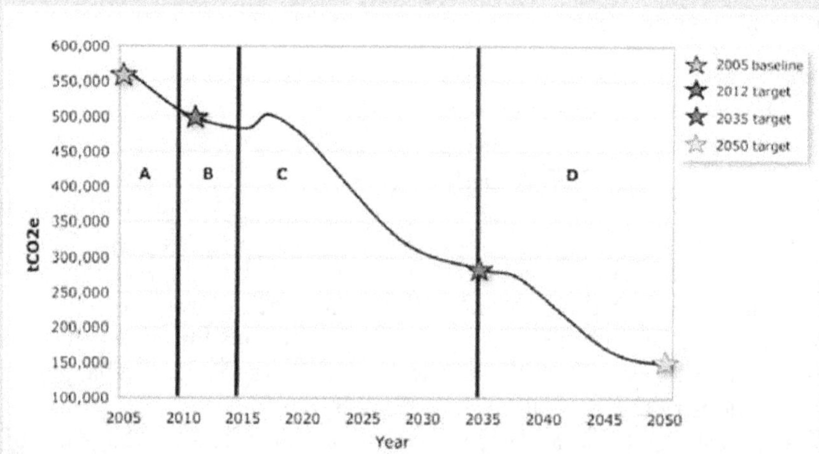

Figure 3.11. United Utilities long-term goals for carbon reduction

Phase A: Setting the agenda 2005–2010

A short-term program of measures has been implemented in AMP4 making reductions to the baseline carbon footprint and that will contribute toward our 2012 target of a gross 8% reduction. (We have purchased REGO certified green energy that is worth a further 18% reduction.) This program consists of projects that apply known technology, including combined heat and power (CHP), pump efficiency, and our first sustainable catchment management program (SCaMP). Collectively these projects will deliver a reduction of 39,000 tCO_2e (based on an electricity conversion factor of 0.43) by 2012.

Phase B: Consolidation and preparation 2010–2015

This second phase of our program will be based on continued implementation of existing technology as well as through introduction of new projects and technologies, such as composting grit and screenings. The program will further harness renewable energy sources with the planned introduction of additional hydroelectric turbines and development of the UK's first biomethane generation plant.

A substantial research and development program is planned that will implement technology targeted at reducing energy consumption and maximizing energy and other valuable resource recovery. This technology will benefit us and our customers as well as the wider society. The outcome of this research will equip us with the knowledge we need to achieve our medium- and longer-term goals.

We will also continue to investigate our operational activities, such as integrated urban drainage, to inform and develop more sustainable strategies.

Phase C: Changing the way we operate 2015–2035

Future opportunities to reduce our carbon footprint will require advances in technology. Therefore, to meet our medium-term goals, we will need to develop and implement the new processes or technologies that we will have developed from our R&D program. Novel ideas will be developed, pilot-tested, and implemented during AMP5. We will quantify the benefit and cost of potential technologies. This third phase of our plan will show the effectiveness of the projects developed over the period 2010-2015, and we will begin to see the start of the most significant period of emission reduction.

We will also actively investigate how we can affect and influence our customers to work with us in reducing our emissions and continue to inform our strategies.

Phase D: Evolution of the water industry 2035–2050

Our commitment to a sustainable future includes playing our part in helping the UK achieve the long-term reduction goals. To achieve this will require a fundamental change in the way we operate. Not only will we need to operate at the lowest possible level of emissions but we will also need to actively encourage and help our customers and other stakeholders to do likewise.

And if appropriate, separate such items as:

- pallets,
- cell phones,
- CDs and DVDs,
- printer and toner cartridges,
- Styrofoam (block packaging and packing peanuts),
- electronic items,
- uniforms, and
- shoes.

Weigh each of your categories and record the weight. This gives you the data needed to calculate the percentage of your garbage each type of recyclable item represents.

You can then come up with a simple chart. The first graph (Figure 3.12) shows the percentage of items, by weight, that was in the trash but could have been in the recycling. It compares two different audits, more than two years apart, and shows that progress has been made overall.

The second chart (Figure 3.13) goes into more detail, showing that most categories of recycling were finding their way into recycling more in 2005 than in 2003, but office paper, of all things, was a problem. Information like that can lead to questions about whether there are new employees who need more education, a janitorial staff who is throwing out items by mistake, or a new process that is generating paper that is getting thrown in the trash.

Charts can be as creative or as simple as you'd like. Basic Excel® pie charts will tell your story, too. Figure 3.14, for instance, shows that of the items cataloged from the trash, 56% of them really belonged in the trash. But around 44% did not. About 9% were junk mail, envelopes, paper bags, newspapers, catalogs, or cardboard. Another 7% were containers—tin and aluminum cans (this

To track and reduce its carbon emissions, TVWD participates in Climate Leaders, a US Environmental Protection Agency (USEPA) industry-government partnership that works with companies to develop long-term comprehensive climate change strategies. Participants set an organization-wide greenhouse gas reduction goal and inventory their emissions of the six major gases (CO_2, CH_4, N_2O, HFCs, PFCs, and SF_6) to measure progress. They then set an aggressive organization-wide GHG emissions reduction goal to be achieved over the following 5 to 10 years. A GHG emissions inventory is completed annually and data reported to the USEPA.

Organizations that emit more than 10,000 tons (9,071.85 metric tons) of CO_2e are eligible to receive free assistance from USEPA to help them determine their emissions and identify ways to reduce them. Other organizations may join USEPA's Low Emitters program that offers simplified tools and Web-based assistance.

The Climate Leaders GHG Inventory Protocol is based on an existing corporate GHG inventory protocol developed by the World Resources Institute (WRI) and the World Business Council for Sustainable Development (WBCSD) as described in the text.

USEPA's GHG calculator (found at: http://www.epa.gov/climateleaders/documents/sgec_tool_v2%208.xls) is valuable even if you do not join the Climate Leaders program.

About two thirds of TVWD's emissions relate to electricity, with about 80% of that electricity usage attributed to pump stations. About 15% of TVWD's emissions stem from fleet usage, 13% from employee commuting (an optional calculation), and about 3% each from business travel (also optional) and natural gas usage.

What is Waste?

- Something we have purchased and reduced to a value of zero or less.
- Actions taken that don't create or add value.
- Unnecessary costs–permits, fees, transportation and storage, disposal costs, safety risks, staff time for items that don't add value.
- A sign of inefficiency and lost revenue.

Figure 3.12. Comparison of recyclables in trash at two waste audits

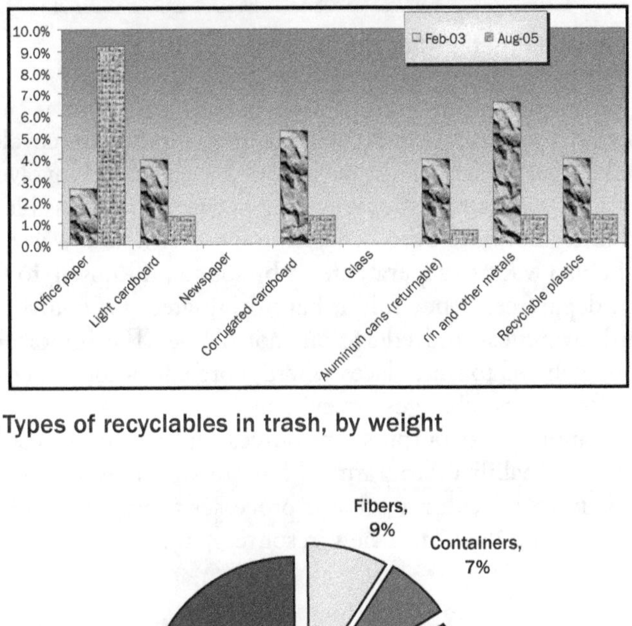

Figure 3.13. Types of recyclables in trash, by weight

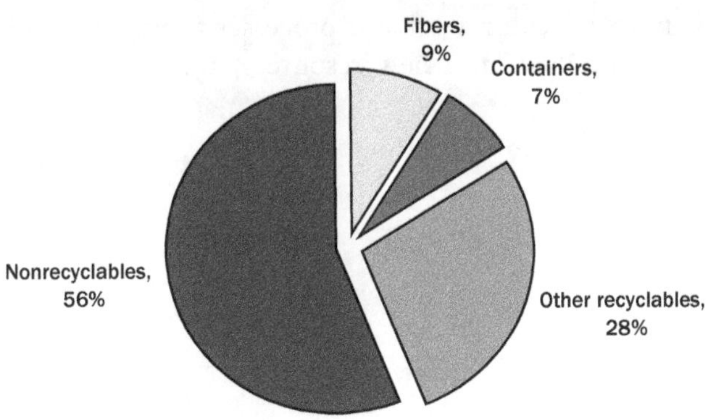

Figure 3.14. Pie chart showing that 44% of items in trash could be recycled

organization discovered during this audit that its aerosol locating paint cans could be recycled), plastic bottles, yogurt-type containers, and milk cartons.

In addition to wood and compostable food, the other main component of "Other Recyclables" in this example turned out to be temporary meter registers (Figure 3.15). These were used in construction areas before the more expensive

Figure 3.15. Unusable meter registers ready for recycling

automated meters were installed. After some investigation, a recycler was found who would take these apart locally, pay a little for the metal involved, and separate out the rest of the materials for recycling of everything possible. This opportunity would likely not have been brought to light without a waste assessment.

You might also want to separate items by source, if known, to find out from which area or department they might have originated. Of course, do this with an eye toward awareness and education, not blame. The importance of waste audits or assessments is to find places where more education is needed or where opportunities for process changes are evident.

Evaluating inputs and outputs—resources and waste—is essential to any water utility's sustainability program. The more we know about the resources we use and the more we can rework our processes to use less initially and turn what waste we do produce into new resources, the closer we are getting to sustainability.

Resources

AXIS Performance Advisors, http://www.axisperformance.com/
- The Business Guide to Sustainability
- The Step-by-Step Guide to Sustainability Planning
- SPaRK (Sustainability Planning and Reporting Kit)
- SCORE (Sustainability Competency and Opportunity Rating and Evaluation) tool

Carbon Emissions Inventories
- Deep Lake Water Cooling System, Enwave Corporation, E.E., http://www.enwave.com/enwave/dlwc/
- Environmental Management System Guidance, http://www.epa.gov/epawaste/inforesources/ems/index.htm
- Greenhouse Gas Calculating Tools, USEPA Climate Leaders, http://www.epa.gov/climateleaders/documents/sgec_tool_v2%208.xls
- Huxley, D., et al. 2009. *Greenhouse Gas Emission Inventory and Management Strategy Guidelines for Water Utilities.* Denver, Colo.: Water Research Foundation.
- International Organization for Standardization ISO 14001 certification, www.iso.org
- Renewable Energy Case Study Metro Hall—Deep Lake Water Cooling, http://www.toronto.ca/ewmo/pdf/dlwc.pdf
- Water Footprints, www.waterfootprint.org, http://virtualwater.eu/

References

Arpke, A., and N. Hutzler. 2006. Domestic Water Use in the United States—A Lifecycle Approach. *Journal of Industrial Ecology,* 10:(1–2); 169–184.

AwwaRF. 2007. *Energy Index Development for Benchmarking Water and Wastewater Utilities.* Denver, Colo.: Water Research Foundation.

Cohen, R., B. Nelson, and G. Wolff. 2004. *Energy Down the Drain: The Hidden Costs of California's Water Supply.* New York: Natural Resources Defense Council.

Foley, L., and P. Lant. 2007. *Fugitive Greenhouse Gas Emissions from Wastewater Systems.* Melbourne: Water Services Association Australia.

Hitchcock, D., and D. Atwood. 2002. *Developing Effective Systems for Managing Sustainability: Creating an EMS-Lite.* Portland, Ore.: AXIS Performance Advisors.

Hitchcock, D., and M. Willard. 2006. *The Business Guide to Sustainability: Practical Strategies and Tools for Organizations.* London, UK: Earthscan.

———. 2008. *The Step-by-Step Guide to Sustainability Planning.* London, UK: Earthscan.

ISO. 2009. www.iso.org. Accessed 11/28/09.

Kelly Services. 2009. *Ethical and Environmental Policies Serve as a Magnet in Attracting Employees According to Kelly Services Global Workplace Survey.* http://ir.kellyservices.com/releasedetail.cfm?ReleaseID=419388. Accessed 1/9/10.

Kenway, S., et al. 2008. *Energy Use in the Provision and Consumption of Urban Water in Australia and New Zealand.* Melbourne: Water Services Association of Australia.

The Natural Step. 2009. Compiled from information at http://www.naturalstep.org/backcasting. Accessed 12/15/09.

Racoviceanu, A., et al. 2007. Life-Cycle Energy Use and Greenhouse Gas Emissions Inventory for Water Treatment Systems. *Journal of Infrastructure Systems,* 13:(4); 261–270.

San Francisco Public Utilities Commission. http://sfwater.org/msc_main.cfm/MC_ID/18/MSC_ID/121. Accessed 1/2/10.

USEPA. 2009. http://www.epa.gov/epawaste/inforesources/ems/basic.htm. Accessed 1/22/10.

WSAA. 2007. *Energy and Greenhouse Mitigation Strategies.* Melbourne: Water Services Association of Australia.

———. 2008. *Sustainability Framework: Methodology for Evaluating the Overall Sustainability of Urban Water Systems.* Water Services Association of Australia Occasional Paper No. 17.

Chapter 4

What Sustainability Opportunities Exist at a Water Utility?

The practical test of an organization's commitment to sustainability is really the active steps its staff takes in support of any sustainable philosophy adopted. This chapter describes a number of projects that have been implemented by water utilities in the United States, Canada, the United Kingdom, and Australia. You will find actions as quick and easy as turning off lights and copying on both sides of the paper. You will also find unique and complex ideas that will require planning, upper level support, and financial investment. I encourage you to put into place anything and everything you find plausible. Then dig deeper to see what might be accomplished with just a little more effort.

At the end of this chapter, you will find a matrix listing almost 200 actions and the department or program likely to be responsible for and affected by the action. Some are extremely simple and you are probably already doing them. Others may not be as easily achievable, but are still possible. Many of the items listed are described in detail below. I hope you find many that will inspire you to make some positive changes in your utility.

Policies And Utility-wide Initiatives

1. Ensure Management Support of Sustainability

Whatever staff are involved in various elements of sustainable operations, the process of making your utility more sustainable will be easier if management and elected officials are solidly behind it. Beyond just memos and tolerance of Green Teams, these folks need to both voice their support to staff and public and behave in ways that back up their words.

A manager who takes public transportation or an elected official who works to reduce his or her own carbon emissions does far more than a proclamation on Earth Day, although those are good, too. In documents, make sure the strength of support comes through, as in these words from England's United Utilities' CEO Philip Green in their Corporate Responsibility Policy:

> **Leadership Responsibility**
> It is the responsibility of managing directors to apply this policy in their areas. Leaders must ensure that their people understand their obligation to manage our economic, environmental and social impacts responsibly. They must also make sure their people have access to and are familiar with this policy and other relevant policies.
>
> **Who Does This Policy Apply to**
> This policy applies to everybody and across all of our business operations where we have direct management control. Where we are involved in joint ventures and where we are working for external clients, we will use our position to influence decisions in line with this policy. This policy is reviewed every two years.

2. Hire a Sustainability Coordinator

While sustainability can be improved without a coordinator, it is very helpful to have a person focused on these endeavors. Sustainability involves changing the status quo, and that is usually met with some resistance. It is important to have a person on staff whose responsibility it is to keep issues in front of management and staff. Exactly what the person focuses on will depend on your utility's priorities and opportunities. But just having someone whose responsibilities include sustainable practices will greatly increase the chances your organization will explore sustainable practices long enough for them to become embedded in your culture. Appendix F lists elements you might want to include in a job description for a sustainability coordinator.

If you are unable to add this full-time position to your staff, see if someone in another position is appropriate to take on some of these tasks. Sometimes a sustainability coordinator ends up in the building and grounds department because the organization is focused on environmentally oriented initiatives best implemented by this department. Sometimes these duties are undertaken by a strategic planner or engineer. Look at your utility's staff structure to see where this position or these duties will be best positioned. In smaller utilities, any environmentally interested staff member might head up a team. This would work best coupled with strong management support.

TIP: The advantage of a full-time sustainability coordinator is that this person can put their full attention to looking at the organization through a sustainability "lens." He or she can fully represent sustainability in organization-wide initiatives. On the other hand, adding sustainability to a current staff member's duties means that you are tapping a person who already has knowledge of the workings of the organization and might be better able to identify who in the organization will be supportive of sustainability and to assess what paths will be most successful in your particular utility.

3. Adopt a Sustainability Resolution, Vision, and/or Principles

It is important that you know where you're going, and a formal vision and principles may help you chart that course. A sustainability resolution (or vision or principles) adopted by your elected officials might be among the first actions you take, but it doesn't have to be. Develop a public expression of commitment to sustainability at some point. But don't feel you have to make this statement before you do anything. Indeed, it may take some successful projects or actions to convince your elected officials that sustainability should be so ingrained in the fabric of your utility that a resolution is needed. See appendix H for a couple of examples.

4. Develop a Sustainable Purchasing Policy

If legally allowable in your area, you may want to give recycled or otherwise sustainable products a price preference, so they can be purchased even if they are 10 or 15% more expensive. That tack can be somewhat limiting, however, as one particular percentage is not always appropriate for all products. Another option is to specify the attributes you want, for example, 100% postconsumer waste recycled content, Green Seal certified cleaning products, hybrid vehicles, and so forth, and take the lowest price for the type of product you want.

See appendix I for examples of sustainable purchasing policies and internal guidance documents. Find out if there are any legal restrictions in your area before you develop your policy.

5. Include Introductory Sustainability Language and Utilize Sustainability Criteria in Requests for Proposals

One of the best things we can do to spread the concept of sustainability in our industry and its supply chain is to request sustainable products and behavior from our suppliers.

You can do this by including introductory language in each request for proposal (RFP) that lets bidders and vendors know sustainability is important to your utility.

Here are some examples:

> Our utility is dedicated to providing our community quality water and customer service in a manner that protects the earth's environment, supports a strong economy and promotes a robust community.
>
> Proposal Format: In keeping with the District's sustainability principles, proposals should be duplex printed on at least 50% recycled paper. No binders, spiral bindings or binding combs are required.
>
> (Tualatin Valley Water District, 2009)

> Sustainable building integrates building materials and methods that promote environmental quality, economic vitality, and social benefit through the design, construction and operation of the built environment. Sustainable design, construction and operations merges sound, environmentally responsible practices into one discipline that looks at

> the environmental, economic and social effects of a building or built project as a whole. These sustainable aspects of the development process encompass the following broad topics: efficient management of energy and water resources, management of material resources and waste, protection of environmental quality, protection of health and indoor environmental quality, reinforcement of natural systems, and integrating the design approach.
>
> For best results, sustainability should be clearly articulated as a guiding principle for project development, and incorporated into the project from the earliest stages.
>
> (Seattle Public Utilities, 2009)
>
> **Procurement Requirements**
> All printed and photocopied documents related to the fulfillment of RFPs and contracts must, whenever practical, use recycled paper, be double-sided, and otherwise fully comply with the provisions of the Environmental Purchasing Policy.
>
> Minimize the number of bound proposal submissions required and encourage electronic submissions for proposals as an alternative to multiple paper copies.
>
> All RFP documents should include information on the City's policies and programs to buy recycled and environmentally preferable products. Vendors should be encouraged to provide submissions on products with recycled-content or which meet criteria for environmentally preferable products whenever such products meet the performance criteria specified in the RFP documents.
>
> (City of Boulder, Colo., 2009)

If appropriate to the bid, ask bidders to describe their own sustainable practices or products. Ask if they have sustainability policies or principles of their own. Assign a percentage of points in the RFP to this.

6. Integrate Sustainability into Regular Business Practices

Look at your processes for budgeting, capital improvement plans (CIPs), strategic planning, purchasing, and investing. Examine current documents and policies to see how you might incorporate sustainability into what you are already doing. The more you are able to do this, instead of adding sustainability as a separate initiative or layer, the better.

7. Add Sustainability Criteria to Capital Improvement Plan Evaluations

Appendix J shows two examples of criteria and weighting systems you might use to evaluate the sustainability of proposed CIP projects. Use this type of evaluation as a source of discussion to give your projects more environmental and social strength, or use it as part of a ranking system to actually choose what projects are completed.

TIP: The precise scores you attach to various projects are not as important as the discussion of how each project performs against your criteria. Your chief engineer, sustainability coordinator, and other key planning and construction staff should be a part of rating these projects. Often projects can even be modified during these discussions to make their construction more sustainable.

8. Establish a Green Information Technology Initiative

Green IT is a set of principles and practices for managing computing technology to improve efficiency, reduce power consumption, and manage hardware purchasing and disposal in environmentally responsible ways. It covers the following steps:

a. Measure the total energy use from computing sources.
b. Pursue data center efficiency and power reduction through server consolidation, virtualization, cooling optimization, and "right-size" provisioning.
c. Reduce desktop power use through power management techniques and desktop virtualization.
d. Manage hardware procurement using environmental and power efficiency standards.
e. Manage electronic waste (e-waste), server decommissioning, and hardware disposal using environmental standards and guidelines.
f. Consider shared computing options to reduce the internal computing footprint, including hosted applications and cloud computing services.
g. Implement computing technology to support energy savings or other sustainability initiatives (sometimes referred to as *Green IT 2.0* or *IT for Green*).

Green IT is a great way for an IT department or division to engage in sustainability. It promotes internal teamwork and gives IT staff a clear way to support larger organizational goals. Data center efficiency is often challenging but can have a significant impact on energy consumption, particularly in larger organizations that have experienced data center expansions. Green IT has other organizational benefits, including cost savings, improved infrastructure manageability, and increased flexibility with business continuity and disaster recovery.

9. Develop Performance Measures That Address Sustainability Issues

If your utility uses performance measures, be sure to include objectives and targets that address such things as energy efficiency and community involvement among your measures. See chapter 5 for further discussion.

10. Develop a Carbon Emission Reduction Policy or Climate Change Response Plan

See discussion in chapter 3 and appendix E.

11. Develop a Bidder Integrity Policy

You may include in your sustainability policy or in a separate document a requirement that vendors for certain types of items live up to specific standards or their bid will not be considered. You might initiate a policy that requires suppliers of pipe, hydrants, fixtures, or other major purchases to be free of current unresolved Occupational Safety and Health Administration (OSHA) and USEPA claims in order to sell to you. This shows vendors that you are paying attention to the way they operate their organizations and that price is not the only consideration. Make sure you use this policy only for products where enough vendors exist to allow you to not purchase from those vendors who do not meet the criteria.

You might also give consideration in requests for proposals (RFPs) to local, in-state, or in-region suppliers, or those meeting specific sustainability criteria as appropriate for the product or service required. Create a preference for buying local and only buy outside of your geographical region if items are not available close to your location. See example in appendix K.

12. Establish Sustainability Teams or Task Forces

Even the most sustainable building is useless unless its occupants behave in sustainable ways. To mobilize and involve employees in sustainability efforts at all levels of the organization, consider establishing two different groups: a sustainability steering committee and a green team, or sustainability team. These two groups can work in tandem with the utility's sustainability coordinator to drive progress toward sustainability at the utility.

The green team might be a grassroots employee effort that educates employees about cost-effectively protecting the environment and benefiting society by their actions. The group brings information and speakers to employees and proposes and encourages policies and practices that minimize or eliminate the utility's impact on the environment.

You may also want a more senior team to address the addition and modification of policies to embrace sustainability. Purchasing, human resources, capital improvement planning, and many other departments and processes will likely be affected by more sustainable actions. The organization may be asked to adopt a sustainability resolution or modify its purchasing policies. Your senior management team may want to take on these decisions, but you also need a representative who was a part of the process.

Ideally, a sustainability steering committee that includes senior management plus several others from throughout the utility would be created to become educated and participate in the process of these modifications. This allows senior managers to understand the background behind requests brought to them. If that is not possible, the sustainability coordinator or other representative should be given time to thoroughly educate the managers about whatever sustainable issue is before them. Seeing activities through a sustainability "lens" may be unfamiliar and will require time and training for all involved.

Decide where your sustainability teams can make the most impact, for example, educating employees, developing proposed policy, and/or implementing and promoting particular projects. Every organization will be different.

Recruit people who are passionate *and* people who are in positions of influence. I actually recommend calling it something other than a green team, as you will want the name to connote more than just the environmental aspect of sustainability.

13. Establish a Carbon-reduction Task Force or Energy Team

You might want to establish teams, at least temporarily, to address particular major issues. A climate change team, for example, would focus specifically on ways to reduce your utility's carbon emissions through electricity, natural gas, fleet, etc. An energy team would take a close look at ways to reduce energy usage. Include on the team people who have the influence to make changes when they are identified. If you feel your sustainability steering committee is adequate, these additional teams may be unnecessary.

14. Go Through a Backcasting or Similar Process to Develop Targets for Sustainable Goals

Backcasting is a planning approach that begins by envisioning desired future states, then works backwards to connect that desired future to the present by integrating policies, programs, and technological innovation. This process is heavily utilized by The Natural Step, and a more detailed description can be found in chapter 3.

15. Undertake a Process Mapping Exercise to Identify Resources Used and Waste Created in Agency Processes

See the sidebar on p. 33 for a detailed description of how to do this.

16. Conduct an Agency-wide Carbon Emissions Inventory

See chapter 3 and appendix E for a detailed description of how to do this.

17. Adjust All or a Portion of Your Carbon Emissions With Renewable Energy Credits and Carbon Offsets

Putting into place actions and equipment that will reduce the amount of greenhouse gas (GHG) emissions your utility is responsible for is the best course. However, since you are not likely to reduce your GHG emissions to zero, purchasing carbon offsets and renewable energy credits (RECs) is an option when your desire is to reach carbon neutrality. RECs are used to offset purchased electricity; carbon offsets are used to offset the rest of the GHG emissions discussed in chapter 3. In both cases, you are essentially purchasing the environmental attributes associated with electricity generation from new renewable technologies like wind and solar energy.

A renewable energy facility produces two distinct "products." The first is electricity. The second is the package of environmental benefits resulting from not generating the same electricity—and emissions—from a conventional gas or coal-fired power plant. The renewable electricity displaces the dirtier power that would otherwise have been generated and delivered to the power pool.

These environmental benefits can be separated from the electrical power and packaged into an offset. The offset then represents the collective environmental benefits of avoided CO_2.

The offsets reflect the avoided pollution that results from substituting new wind, solar, and other renewable electricity sources for coal or gas. The electricity generated by the renewable project is sold as generic power, separately from the carbon offset. No environmental claims can be made for this power because the carbon offset now represents the entire package of environmental benefits associated with these specific megawatt hours. The distinction between the two products is protected by enforceable contractual agreements (Bonneville Environmental Foundation, 2009).

While you will not be getting the precise electrons generated by the renewable power you purchase, this renewable power is mandated to replace petroleum-based energy on the grid, so it does change the energy mix as a whole. There now are many providers of both products, so be sure to research who you will purchase from, or send out an RFP so you are sure you are getting what you want.

18. Join a Sustainable Networking Group

Join The Natural Step or other group of people working to improve their organizations' sustainability. Whether a recognized nation-wide organization or an informal group of like-minded people trying to create change in a particular community or industry, it is very helpful to have others to network with. Sharing successes and challenges, within or across industries, allows sustainability to progress faster for all involved. People in this field tend to be very willing to share practices, documents, art work—anything to make it easier for the next person to be successful. There seems to be a shared understanding that we are all in this effort together, and the more we can build on each other's work, the better.

Contact The Natural Step, search the Internet for groups in your area, or check the phone book or community bulletin boards. If nothing exists in your area, you have an opportunity to develop something. AWWA Water Resources or Conservation committees at your section or subsection level might be good places to find others interested in making their utilities more sustainable.

An international organization, the International Society of Sustainability Professionals (ISSP), is another good resource. It is a nonprofit, member-driven association for professionals committed to making sustainability standard practice. Members share resources and best practices and develop themselves professionally. Special reports, salary surveys, and networking phone calls are some of the benefits offered members (www. sustainabilityprofessionals.org).

19. Develop a Sustainable Food/Catering Policy

Research establishments in your area that are working toward being sustainable in that industry. This might involve serving seasonal, locally grown organic food; it might mean serving meals family-style in large serving containers, then returning to pick them up. A sustainable caterer might bring reusable plates,

> TIP: Incorporating sustainability into human resources policies will take some training. Employees will need to get used to noticing and talking about their sustainable efforts, and supervisors will need to be trained to include that in their evaluations.

cutlery, glasses, and even napkins. See appendix I.2 for a list of signs that you have found a sustainable caterer. Don't be afraid to ask potential vendors about some of these things—suppliers will change their practices more rapidly if customers are asking these kinds of questions.

At the very least, when you order food from an outside source, ask for it to be delivered in non-single serving containers. See "72. Serve 'Family-style' Meeting Meals Instead of Individual Servings" under Sustainable Purchasing.

20. Incorporate Sustainability Into Human Resources Policies

Incorporate sustainable language into your job announcements. Simply saying, "We are a sustainably oriented employer" will get your message across.

Add sustainability aspects to performance reviews. You might choose to add a specific section for sustainability-related actions. The opportunities available to each employee will differ based on the specific job, though. One solution is to look at various categories of positions, for example, office workers, engineers, meter readers, maintenance crews, construction crews, etc., to see what specific opportunities might apply to each category. Some examples of ways for employees to excel in their jobs with regard to sustainability are shown in appendix G.

21. Create an Employee Recycling Benefit

Do anything you can do to make sustainable actions easier than less sustainable ones. If employees are able to bring items to work for recycling, for example, they are far more apt to actually recycle them. Given the chance, most people want to see items at the end of their usable lives go to others who can use them or be made into new products.

Energy Issues/Carbon Emission Reduction

22. Use Natural Light and Occupancy Sensors Where Appropriate

Natural light, occupancy sensors, and lighting control panels will help conserve energy by shutting off lighting during periods of nonuse and by reducing the amount of exterior lighting needed during evening hours and weekends. Natural lighting has also been shown to improve productivity and reduce errors, as well as improve staff morale.

A brainstorm of its human resources manager, TVWD employees are now able to bring in items for recycling that they are unable to recycle at home through curbside pickup. As public employees, District employees may not receive any benefits that are not also available to the public. Therefore, TVWD could not allow employees to bring in recyclables without taking them from the public also, which District facilities were not equipped for.

By creating an employee recycling benefit as part of the staff benefit package, workers may now bring in everything from batteries to electronics to tennis shoes, and the items are sent out for reuse when possible, or at least recycling. An employee recycling center was established to provide a convenient outside drop-off location for employees as they arrived for work. The TVWD Board of Commissioners was fully supportive of this plan, approving it unanimously in September 2005. The only cost is for recycling some electronics, which the District covers for employees. All items recycled are sent to reputable collectors who keep as much material as possible within US boundaries. Staff costs are not increased because each type of item would have been taken in for reuse or recycling anyway, as the District accumulated Styrofoam, pallets, and other items.

23. Participate in an Electricity Provider Program That Shows Your Electricity Usage in Small Intervals

Some electric utilities can install equipment that will give you readings from your meter every 15 minutes or so. This is followed up with monthly reports that allow you to see anomalies and investigate and correct reasons for excess usage. This information will help you discover opportunities for electricity reduction.

24. Install a Programmable HVAC System

Building occupancy times that are precisely programmed into heating, ventilation, and air conditioning (HVAC) equipment with a digital control system will conserve energy by shutting down heating and cooling equipment evenings, weekends, and holidays. Couple this with the electricity provider program above to quickly discover and fix problems in the settings.

25. Turn Off Lights, Computers, and Electronic Devices When Not in Use

First, make sure equipment and lights are not left on when not in use. That's the easy part.

More insidious is the "vampire" or "phantom" load, which refers to the electric power consumed by electronic appliances while they are switched off or in a standby mode. Each hard drive could use 310 kW•h/year; each monitor, 23; and each laptop, 145. Depending on your power costs, that could be \$15–\$35 per year per work station.

Educating your staff will help them at home, too, as TVs, cell phone chargers, almost anything electronic can suck power even when the item appears to be off. While VCRs and DVD players might cost an employee \$5–\$10 a year

TIP: Seeing progress is a good thing. Create a basic graph of electricity usage and post monthly usage on employee bulletin boards so they can see trends over time. Combined with increased awareness and education about issues like "vampire" energy, you will likely see considerable reduction in energy usage. This works best in facilities where usage is relatively stable. Most pump stations, for instance, would be more difficult to analyze, as there is more fluctuation in usage.

(75–90 kW•h), a plasma TV in active standby mode could result in electricity charges upwards of $150 a year (1400+ kW•h) (GOOD Web site, 2009).

Power monitors (Kill-a-Watt™, Watt's Up Pro™, and more sophisticated models) will calculate this usage and prove that electronics can use amazing amounts of electricity and money with no resulting value. To counteract this, use power strips to handle desktop items such as hard drive, monitor, desk lamp, and printer and encourage staff to shut power strips off when they leave.

Distribute physical prompts (cards to attach to computers or put next to light switches) and e-mail reminders to employees. Make sure prompts are colorful with catchy phrases so they are not just overlooked.

26. Reduce Desktop Computer Power Consumption

Up to 50% of the power that typical office computers use is wasted. Newer operating systems (Microsoft Vista®, Microsoft Windows 7®, and Mac OS X®) come with advanced power management capability. Use these power management settings to let inactive computers go into sleep or hibernate mode. Turn off screen savers—these often cause computers to draw more power to display animation or rotating images.

There is also software available to manage personal computer (PC) power status centrally. Tools such as Surveyor by Verdium and BigFix Power Management support network-wide solutions for desktop and laptop computers. Shutdowns can be automated and scheduled but still allow for remote repowering for off-hours use or to allow for system patch updates. Power Save by Faronics automatically detects a computer's total activity and has configurable settings to control a variety of power-saving actions.

27. Use Renewable Energy

Though solar installations often have a long payback time at present, power purchase agreements (PPAs) make solar power possible in many US locations. Power purchase agreements allow an organization to offer their roof to a third party who installs, maintains, and owns the system. The organization pays the third party for the solar power generated instead of paying their usual electricity provider. This usually requires little or no investment and displaces petroleum-based power from the grid in favor of renewable energy.

New Jersey American Water (NJAW) invested in solar technology for its Canal Road Water Treatment plant in Somerset, N.J. NJAW serves more than 2 million New Jersey residents. To curtail its escalating power-generation costs and follow through on its commitment to preserving the environment, they installed a ground-mounted, dual-array photovoltaic (PV) system at its Canal Road facility in late 2005. At the time, the solar PV system was the largest ground-mounted system on the East Coast (Wolan et al., 2007).

Some of the financial incentives that were available to the utility at the time of construction are detailed here. Check with your state Department of Energy for information on what is available to you.

Financial Incentives Bolster Green Benefits

Along with providing New Jersey American Water with a financial hedge against rising fuel costs, the solar photovoltaic (PV) system at its Canal Road Water Treatment Plant is enabling the company to secure a strong return on its renewable energy investment through financial incentives, such as rebates and tax credits, provided by the New Jersey Clean Energy Program and the federal government.

New Jersey Clean Energy Program. New Jersey's 1999 electricity restructuring legislation provides for investments in energy efficiency and renewable energy through the "Societal Benefits Charge" collected from all electric public utility customers. In March 2001, the NJ Board of Public Utilities approved funding for renewable energy programs, including a customer-sited renewable rebate program administered by the state's utilities.

Solar PV systems are eligible for incentives, which are paid incrementally based on the size of the system installed, up to 700 kWdc—for rebates secured after July 1, 2005. At the time NJAW secured a rebate for the Canal Road project, the limit was 500 kW. The company secured a $1.9 million rebate for the project, which was paid at the end of the construction period.

Modified Accelerated Cost Recovery System. Under the MACRS, businesses can recover investments in solar, wind, and geothermal property through depreciation deductions. The MACRS establishes a set of class lives for various types of property, ranging from three to 50 years, over which the property may be depreciated. For solar, wind, and geothermal property placed in service after 1986, the current MACRS property class is five years.

Federal Solar Investment Tax Credit. The federal investment tax credit is a 30 percent tax credit available to commercial businesses that invest in or purchase energy property in the US for systems that are put in service in 2006 and 2007. This incentive was 10 percent for the Canal Road project, which started operation in October 2005.

New Jersey State Sales Tax Solar Energy Systems Exemption. New Jersey offers a full exemption from the 6 percent state sales tax for all solar and wind equipment.

Reduced Electricity Bills (Avoided Costs). The system is expected to produce 13.2 million kWh of electricity over the 25-yr evaluation period. Electricity from the grid was conservatively forecast to increase 4 percent per year. The fixed price for the power from the Canyon Road solar generation system is an effective hedge against future price increases, which are a function of volatility in fossil fuel commodity markets.

The avoided cost of electricity is estimated to be $0.09 per kWh. The avoided cost of electricity for PV systems generating in New Jersey, on a real-time basis, has been analyzed to be 18 percent greater in value than average power.

Renewable Energy Credits. The State of New Jersey Clean Energy Program has created a market for tradable RECs for purchase by electricity suppliers that supply end-use customers located in the state. Suppliers must meet the Renewable Portfolio Standard, which specifies the percentage of electrical energy, measured in kWh, that must be generated by a renewable source, such as solar or wind. Suppliers that fail to produce or secure RECs will be subject to punitive fines for noncompliance.

The Canal Road financial analysis assumed a sustained current market value of $0.15/kWh over the life of the system, which has been verified in the marketplace. The combination of reduced energy costs and sale of solar RECs is estimated to provide NJAW annual savings of about $152,000.

The Portland (Ore.) Water Bureau installed a 267-kilowatt solar array at the bureau's groundwater pump station. At peak generation, the panels produce 267,000 watts of electricity, enough to power about 50–60 average homes in the Pacific Northwest.

Pumping groundwater is the bureau's single-highest use of energy. Installing the solar array at the groundwater pump station helps to offset some of that energy.

Contributed by Kim Dinan, Portland Water Bureau

TVWD's Grabhorn aquifer storage and recovery (ASR) wellhead was at an elevation of 396 ft. The adjacent Grabhorn Reservoir had an overflow elevation of 385 ft. With the system configuration, however, there was no direct method for injecting water into the well from the Grabhorn Reservoir or the 385-ft pressure zone.

Water was first pumped from the 385-ft pressure zone to the 550-ft pressure zone. That water was pumped again to a hydraulic grade line (HGL) of 800 ft by the Goyak Pump Station. The 800-ft pressure zone water was conveyed across the 800-ft pressure zone to a pressure regulator in Grabhorn Road where it was reduced back down to 550 ft HGL. The 550-ft pressure zone water underwent further pressure reduction before being injected into the aquifer at an elevation of 396 ft. Extensive energy was needed and the system was very inefficient.

Previous injection cycles had also shown the aquifer to be sensitive to turbid water. Turbidity was variable from the district's east-side supply, depending on season, and typically high in the winter months. Water from the west-side supplier consistently had a much lower turbidity level, making it the water source of choice for injecting into the well. Isolation of the west-side source was possible through isolation of a transmission main from the 75th Street Fluoride Station to the 189th Street Pumping Station.

Three options were identified for delivering water to the well:

- Operate the system as is.
- Install a low-head transfer pump.
- Deliver high-pressure, west-side water to the well.

Operate the system as is:

This option would continue to reduce pressure at the 75th Street Station to be pumped twice again before being reduced two more times prior to being injected into the well. This option would consume 600,000–650,000 kW·h per annual six-month injection cycle and did not align with the sustainability goals of TVWD.

Install a low-head transfer pump:

This option would include installing a 30-hp submersible pump in the Grabhorn Reservoir. The pump would transfer water directly out of the reservoir into the well. This option would also include a well discharge line that would allow water to flow from the well into the reservoir when water was pumped out of the well. Construction of the low-head pump and return line would be difficult and expensive. Although the pump would be a much more efficient way of injecting water than the current operations, it was not the most sustainable option. The pump would operate for six months of the year and consume approximately 64,000 kW·h per year. The pump would create maintenance and possible security risks with a portion of pipe running from the reservoir to the

(continued)

existing dirt bank surrounding the reservoir. In the event the pump needed maintenance or repair, the reservoir would have to be drained or a diver would have to be sent in.

Deliver high-pressure, west-side water to the well (chosen option):

TVWD has a connection to the west-side South Transmission Main at SW 75th St. The west-side water enters the 75th Street Station at an HGL of 520 ft from the Fern Hill Reservoir. The water is gravity fed to the 75th Street Station where it is reduced to a 385 ft HGL. This option involves increasing the regulated pressure at the 75th Street Station to allow the higher incoming pressure to transmit water through the isolated transmission main. The increased pressure is enough to reach the ASR well without pumping, thus consuming no electrical pumping energy.

The system will be operated in this isolated manner during the ASR injection cycle (winter months) and returned to normal operating conditions when pumping from the well (summer months). The system has historically been operated in the "isolated transmission main" mode during injection (though at a reduced pressure, intended to direct west-side water to the 189th Street Pump Station). Based on hydrant flow testing at four locations, the system provides adequate fire flow while operating under this isolated condition.

This option required the installation of approximately 900 ft of 12-in. ductile iron pipe. With a focus on a sustainable future and an emphasis on safety for employees, this option is the lowest-cost alternative, involves the least maintenance and risk, and will provide a long-term benefit with an upcoming regional water supply project.

The new pipe and modified distribution system has performed as planned. Initial results show that a small portion of the water that was being injected into the well actually came from the 920-ft pressure zone, meaning that it had been pumped a third time and to a HGL that is more than double what is actually required for injection.

Contributed by Ryan Smith, Tualatin Valley Water District

28. Use Gravity Instead of Pumping to Move Water

Whenever you can, look at infrastructure through a sustainability lens. How can you use less energy to move water? One situation where gravity saved a utility significant kilowatt-hours, dollars, and carbon emissions is described on p. 67.

29. Install Variable Frequency Drives Where Appropriate

Pump stations, which bring water to customers at higher elevations, may be one of your biggest energy consumers. Evaluate and install variable frequency drives (VFDs) at pump stations as appropriate for more efficient electricity use.

A variable frequency drive is an electronic controller that adjusts the speed of an electric motor by modulating the power being delivered. Variable frequency drives provide continuous control, matching motor speed to the specific demands of the work being performed. They reduce electrical and mechanical stress on the systems, which can result in fewer repairs and reduced maintenance. VFDs use less electricity than conventional drives and are sometimes covered by utility incentives or rebates, providing an even more rapid payback.

Ten percent or more of a utility's operating budget is spent on energy, and nearly all of that on pumping. The Department of Energy estimates that 75% of pump systems are oversized, many by as much as 20%. Oversized pump systems waste energy because the pumps do not operate under the conditions that they were designed to be most efficient. In some instances, the oversizing of pumps is compensated for by "throttling" or partially closing valves. Upgrading systems with high-efficiency pumps and motors selected to meet current demands frequently provides a quick return on investment. A utility may further optimize the performance and efficiency of their pump systems by installing variable frequency drives (VFDs) on existing or upgraded pumps.

Pump efficiency changes with operating conditions. Well-matched pump systems operating under conditions that do not vary significantly during the day, or from season to season, may consistently operate near their optimal efficiency. Systems operating over a range of conditions do so at higher or lower efficiency. VFDs provide pump systems with the ability to operate closer to their optimal efficiency over a wider range of conditions. For a 50-hp pump system, an average gain of 2% in pump efficiency provides an annual savings of $800 and reduction of 5 tons of indirect CO_2 emissions.

Optimization of pump systems requires flow and pressure data and an understanding of the utility's operations. Where available, supervisory control and data acquisition (SCADA) systems are a valuable source of these data. Flow and pressure data are used to analyze combinations of pumps and VFDs and identify the configurations that provide the greatest benefits. After upgrades, a SCADA system can also be used to optimize operation of the VFD-equipped pumps and to quantify energy savings.

Because VFDs have limitations and are not appropriate for all applications, it is important that the application be reviewed by a qualified specialist. Issues that need to be considered include limited range of flow rates, inadequate rating of older motors, and vibration. In general, systems with high static head benefit less from VFDs.

Contributed by Daniel Haddock, Layne Christensen Company

30. Optimize Pumping Systems

By some estimates, 90% of a utility's energy consumption is for moving water. Important pumping efficiency gains may be achieved with high-efficiency pumps and motors carefully matched to demands. Proactive maintenance and elimination of energy-wasting losses caused by undersized piping and throttled valves are also important. A significant portion of the energy used for moving water is required to lift it from the elevation of the source. Groundwater systems can reduce the depth from which water has to be lifted by optimizing the pumping of wells.

When pumped, the water level in a well drops from the natural "static" level in the aquifer to a lower "pumping" level. The difference between the static and pumping

Finding more efficient ways to move water, either through sustainable energy sources or more efficient pumps, motors, and operations, is where the big gains will be. Reducing per capita demands through conservation is the first step, because that saves both water and energy.

Douglas Yoder, Miami-Dade Water and Sewer Department

Most of Portland's water distribution system was designed to allow gravity to move water from the Bull Run Watershed to the homes of over 800,000 Oregonians. However, some pumping is required to deliver water to parts of the city.

Knowing that pumping water is an energy-intensive task, bureau engineers evaluated the pumping efficiency at a location where two pump stations delivered water to the same pressure zone. The engineers modified the pumping scheme to optimize the gallons of water pumped per kilowatt-hour of electricity used.

The overall efficiency gain was 17%. In the first year, the bureau saved $67,000 in electricity costs. The energy savings is enough to power 90 Oregon homes annually and to eliminate 560 tons of carbon dioxide emissions into the atmosphere.

Contributed by Kim Dinan, Portland Water Bureau

level is the drawdown. The harder the well is pumped, the greater the drawdown and the greater the energy required to lift the water to the surface. Typically, a utility will pump all of its wells only a few weeks or months of each year.

Most of the year, one or more wells are turned off to match supply to demand. During this period, energy can be saved by managing the wells differently to reduce drawdown. In Figure 4.1, three wells of equal capacity are shown. In the first instance, two wells are pumped at their maximum rate, and in the second, the other three wells are each pumped at two thirds of their maximum rate. The same total amount of water is pumped with less drawdown, requiring less energy. For 1 mgd of pumping, an average reduction of 10 ft of drawdown saves $1,500 per year and avoids 10 tons of indirect CO_2 emissions.

Optimizing groundwater withdrawals may be accomplished with the installation of VFDs that allow pump rates to be controlled. Because VFDs have limitations and are not appropriate for all applications, it is important that any potential application be reviewed by a qualified specialist. Issues that may need to be considered include limited range of flow rates, inadequate rating of older motors, and vibration. In general, systems with high static head benefit less from VFDs. The characteristics of the aquifer must also be understood to predict pumping levels at different flow rates and the potential for optimization.

For larger groundwater supplies, alternative well construction techniques can provide benefits in terms of reduced drawdown and resistance to declining water levels caused by drought or overpumping. In comparison to standard vertical wells, angle wells and collector wells are able to more efficiently capture water in many settings. (This section on optimizing pump systems was contributed by Daniel Haddock, Layne Christensen Company.)

31. Purchase Low-emission, Two-stroke Equipment

The amount of carbon emitted from trimmers, leaf blowers, and similar equipment is nowhere near that from vehicles, but this type of equipment still affects local air quality. It is worth investigating lower emitting equipment when you can. Some manufacturers have an air emissions tag that shows how the equipment rates in air emissions compared to other models.

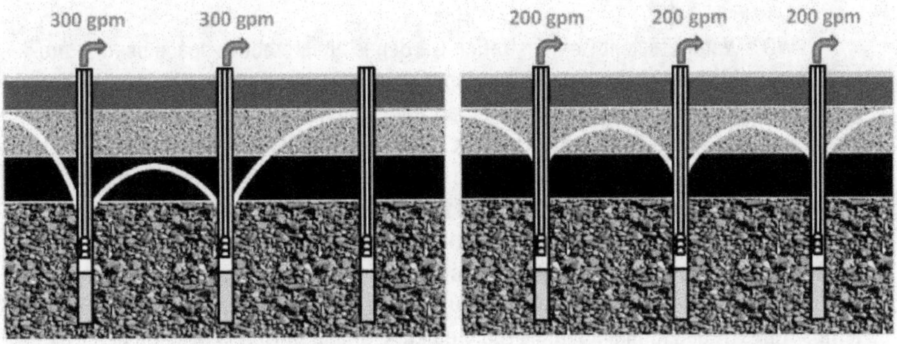

Figure 4.1. Groundwater optimization results in less drawdown, less energy

32. Implement a "No Mow" Program at Appropriate Sites

You may have a long list of remote or not-so-remote sites that were planted in turf long ago and have been kept mowed and groomed ever since. Evaluate your sites to see if there are portions that could be planted with native grasses and brought to a condition that is aesthetically pleasing but does not require mowing. This will reduce staff time, vehicles on the road transporting mowers, and equipment emissions. It may also alleviate some safety concerns if the mowed terrain is somewhat steep.

If the area is large or requires labor to replant, enlist the help of volunteer community members to develop the site and provide labor. This will ensure that the area becomes something nearby residents want and cuts development costs. Add signage to let the public know what is happening and how it benefits the community.

33. Use Goats or Other Animals to Keep Remote Sites Trimmed

Yes, seriously. Partner with a local 4-H group, lease the animals, or buy them outright if you have enough property. Property must be well enclosed, of course, but this can reduce risk of staff injury on steep terrain and reduce carbon emissions and maintenance time associated with trimmers and mowers.

Be sure to consider the health of the animals, too. Perhaps you can partner with local farmers to "rotate" animals to different sites to keep them from overgrazing.

Water Efficiency Within Utility Facilities

34. Install Half-flush or Dual-flush Toilets, Waterless Urinals

Plumbing codes in some areas do not allow these fixtures, but as these codes change, investigate the possibility of installing them in new buildings or when

TVWD's Water Efficient Demonstration Garden is an interactive space used to promote efficient use of water in the landscape. The District remodeled the 2-acre garden site in 2008 to provide a contemporary example of how using water-efficient landscaping and irrigation techniques can result in a beautiful landscape while minimizing excess water used for irrigation and reducing runoff.

The Water Efficient Demonstration Garden (Figure 4.2) provides examples of practical water conservation devices and sustainable landscape practices, such as weather-based irrigation control, drip irrigation, and organic soil amendment. Customers can visit the garden to learn about landscape and irrigation practices that can provide long-term savings, reduced maintenance requirements, and improved landscape health.

The garden design arranges plants into communities of similar water, light, soil, and maintenance requirements. This water-efficient design concept, known as hydrozoning, incorporates native and naturalized noninvasive species into the garden palette.

The garden provides the utility with an "outdoor classroom" to facilitate educational workshops and clinics. Numerous pathways and trails allow visitors to readily move about in order to see, smell, and touch a wide variety of native plants, shrubs, trees, and other landscape elements.

The garden provides food, water, and shelter opportunities for local wildlife and is certified as a "backyard wildlife habitat" by the National Wildlife Federation.

Contributed by Steve Carper, Tualatin Valley Water District

Figure 4.2. TVWD demonstration garden

TIP: Hands-free automatic flushing sensors may negate the savings from low-flush toilets if they tend to flush more than needed—with door or other movement, for example. Test several models to make sure you get what you need.

remodels occur. These fixtures will reduce your water usage and be a model for other organizations. Be sure to put up educational signage to explain the fixtures and the water they save.

35. Use Low-flow Faucet Aerators and Showerheads

Any of your facilities that include faucets and showers should be fitted with low-flow devices. If we ask it of our customers, we need to lead the way.

36. Develop Water-efficient Landscaping

Incorporate the following practices into your garden and grounds:

- planning and design,
- appropriate plant selection,
- practical turf areas,
- efficient irrigation,
- use of mulches, and
- appropriate maintenance.

These measures minimize the amount of water, energy, and labor needed to keep the grounds looking well-kept and professional. If possible, use your own landscaping to demonstrate to neighbors and other interested citizens measures they can take at their own homes to use resources efficiently.

37. Use Weather-based Irrigation Systems

Weather-based irrigation controllers use local or regional weather and landscape conditions to calculate irrigation schedules according to real conditions or historical weather data. Rather than irrigating on a preset schedule, weather-based irrigation controllers are continually adjusted automatically to more closely satisfy the actual water requirement of plants.

Rainwater is captured from the new 6,500 ft^2 (603.87 m^2) wing of TVWD's roof surface, filtered for debris, and stored in a 40,000-gallon (151.42-kL) membrane storage tank located beneath one of our employee parking lots. The rainwater is then pumped through filters and an ultraviolet sterilization system to an isolated piping system that supplies water for flushing toilets and urinals. This reduces the amount of potable water required for flushing by 60% or more. This water is also available for irrigation, if needed. This system was somewhat expensive, but was an investment we chose to make to showcase this water-saving opportunity.

38. Install a Rainwater Collection and Storage System

Explore the possibility of utilizing the rain that falls on your facility before it goes into the stormwater system. Irrigating vegetation and flushing toilets can be successfully accomplished with nontreated water, better utilizing the resource and saving the cost and time associated with treating this water. This is an action that requires you to look at the triple bottom line—economic, community, and environmental benefits—and consider more than just the financial expense of such a system. Installing such a system can show your community that you value water as an important resource and are willing to make an investment to get the most out of it.

39. Employ Stormwater Solutions That Keep Water Out of Storm Drains

Detain stormwater on-site by using an oversized detention pipe or permeable collection areas, rain gardens, green roofs, permeable or porous pavements, and so forth. See http://www.tualatinriverkeepers.org/connect_the_drops.html for many more ideas, as well as descriptions and photos of actual projects.

Low impact development (LID) uses a variety of techniques to control stormwater runoff at the source, instead of relying solely on conventional collection, conveyance, storage, and treatment systems. Techniques include rain gardens or bioretention cells, cisterns and rain barrels, green roofs, permeable or porous pavements, vegetated swales, smart site design, green streets, and reduced pervious footprint. See glossary for descriptions of these techniques.

40. Drain Tanks Through Distribution System

Prior to cleaning tanks, allow water to be "used up," rather than sending it through sewer or stormwater systems.

41. Appropriately Dispose of Water From Continuous Water Quality Analyzers

For those analyzers that don't add chemicals to the water, evaluate the opportunities for the water to do some good. Is there a park or other public facility close by that could use it or is there a possibility of bringing it back into the system? Proactively look for positive ways to use the water before sending it down the drain.

42. Eliminate Oversized Tanks and Other Inappropriate Infrastructure

Reassess your infrastructure to see what problems might be caused by aging infrastructure that is no longer appropriate for your current demographics. For instance, oversized tanks may be causing you to add more chemicals or move more water than necessary simply to keep it clean. As practical, remove such items from use or modify them to be most efficient for their current usage.

Sustainable Construction in New Utility Facilities

When constructing new facilities, guidelines exist to help you build sustainability into a new facility. The Leadership in Energy and Environmental Design (LEED) rating system, for example, addresses six major areas:

- Sustainable sites
- Water efficiency
- Energy and atmosphere
- Materials and resources
- Indoor environmental quality
- Innovation and design process

The system gives a project points for various achievements and assigns a level (certified, silver, gold, or platinum) based on points achieved. Aim to achieve at least silver (gold or platinum is better!) LEED certification or equivalent. There is a fair amount of paperwork involved, but even if you choose not to get the official certification, include elements of this system in your project. Some possibilities:

43. Achieve at Least Silver LEED Certification or Equivalent

44. Reduce Heat Islands to Minimize Impact on Human and Wildlife Habitat

45. Eliminate Light Escaping From Building Site to Reduce Impact on Nocturnal Environment

46. Achieve Increasing Levels of Energy Performance Above Basic Building Standards

47. Include Commissioning to Ensure Building was Designed, Constructed, and Calibrated to Operate as Intended

48. Recycle or Reuse Maximum Construction Waste Possible—90% or More by Weight

49. Use Recycled or Rapidly Renewable Products Whenever Possible

50. Use Materials Manufactured Locally Whenever Possible

51. Use Low-emitting Materials such as Carpet, Paint, Adhesives, and Composite Wood

52. Ensure High Indoor Air Quality Through Construction Process and Monitoring

53. Install Room Motion Sensors or Post Prompts by Light Switches and Computers Encouraging Them to be Shut Off When Not in Use

The Living Building Challenge goes even further than LEED. Its mission: To encourage the creation of living buildings, sites, and communities in countries around the world while inspiring, educating, and motivating a global audience about the need for fundamental and transformative change.

The International Living Building Institute (ILBI) of Seattle, Wash., is a nongovernmental organization (NGO) dedicated to the creation of a truly sustainable built environment in all countries around the world. The Living Building Challenge is a program initially launched and operated by the Cascadia Region Green Building Council (a chapter of both the US Green Building Council and Canada Green Building Council). The Living Building Challenge is comprised of seven performance areas, or "petals": site, water, energy, health, materials, equity, and beauty. Petals are subdivided into a total of 20 imperatives, each of which focuses on a specific sphere of influence (ILBI, 2009).

The Living Building Challenge is designed to be the ultimate in sustainable design, but you can start much more simply. Sustainable construction might include elements such as the following:

Kitchen and break rooms could use low-flow faucet aerators to minimize water usage, Energy Star appliances to save energy and water, cloth dish towels, recycled-content paper towels, and sensor lights. If you have a dishwasher, run it only when full, washing reusable plates, cups, glasses, bowls, and flatware for future use. These rooms should also house recycling and composting containers for ease of use by employees.

Restrooms, too, should make use of faucet aerators, as well as WaterSense rated toilets, urinals, faucets, and showerheads. See Sustainable Purchasing section below for more information on WaterSense fixtures. Bathrooms could also sport dual-flush toilets, enabling employees and customers to use as little as 0.8 gal per flush. Posting educational signs lets people know how much difference such fixtures can make in reducing water usage. Add high-recycled-content toilet paper and paper towels, and sensor lights (or signs reminding staff to turn off lights) to round out the picture.

Outside, select plants and grasses for their drought resistance, compatibility with your local climate, and color throughout the growing season. Then pay attention to fertilizer and pesticides. By using plant material suitable for your local environment and employing weather-based irrigation technology, minimal fertilizing is required.

54. Use Integrated Pest Management

The more you can do to ensure a pest-free landscape without using toxic chemicals, the better. Look first at solutions that are natural, resorting to pesticides only when other actions have proved ineffective.

Integrated Pest Management

Integrated pest management (IPM) is an environmentally sensitive approach to pest control based on an understanding of the life cycles of pests and their interaction with the environment. This information is used to manage pest damage by the most cost-effective means and with the least possible hazard to people, property, and the environment (USEPA, 2009).

Landscaped areas maintained by many water utilities present excellent opportunities to reduce overall environmental chemical load and, with proper plant selection, can also serve as positive examples of environmental stewardship. Consider introducing beneficial predatory insects, such as praying mantis, dragonflies, lady bugs, or even spiders, to provide a safer alternative to chemical pest control.

An IPM system is designed around four components:

1. Setting Action Thresholds

Before taking control measures, IPM programs establish acceptable pest levels called action thresholds. Knowledge of individual pests and a commonsense approach based on site requirements will define the level of acceptable damage or the risk posed by a particular pest.

2. Prevention

Maintaining healthy plants is the first line of defense in an effective IPM program. Selecting pest- and disease-resistant varieties adapted to local growing conditions, as well as proper irrigation, fertilization, and mulching, will serve to produce a pest-resistant plant. These preventative methods can be very effective and present little or no risk to people or the environment.

3. Monitoring and Identifying Pests

Regular observation is vital to the success of an IPM program. Control measures are both more cost-effective and successful when actions are taken at the most appropriate time during a pest's life cycle.

4. Control

Should a pest reach an action threshold, then the least risky (yet adequately effective) control method is first selected. Mechanical control methods may include hand-picking, erecting insect barriers, using traps, vacuuming, or tillage to disrupt breeding cycles. Biological controls include Bt (*Bacillus thuringiensis*) and pheromones that disrupt pest mating, as well as select species of fungi and nematodes that target pests. Many pesticides now available are derived from plants or naturally occurring substances like nicotine, pyrethrum, and insect juvenile hormone. If further monitoring indicates that less risky controls are not working, then additional control methods could be employed, such as targeted spraying of pesticides (USEPA, 2009).

Compiled by Collin Fleming, Tualatin Valley Water District

55. Consider LEED Certification for Water Treatment Facilities

LEED certification is not reserved for office buildings alone. Any structure may be constructed using more sustainable methods and materials, as in the LEED-certified water treatment plants described on p. 78.

In late 2005, American Water began to explore opportunities for enhancing the sustainability of new facilities. The company decided to pursue LEED (NC 2.2) certification for four new treatment facilities constructed in Indiana and Illinois. The four facilities are the Champaign County Water Treatment Plant in Champaign, Ill.; Happy Hollow and Davis Ferry water treatment plants in West Lafayette, Ind.; and the London Road Water Treatment Plant in Shelby County, Ind. The Champaign plant achieved certification in December 2009 and certification is pending for the plants in Indiana.

Some of the measures required for certification were already common practice, while others were given serious consideration for the first time. Measures incorporated into the design and construction of the facilities included site design and pervious pavement (where appropriatc) to reduce runoff, reflective roofing and pavement, native landscaping and prairie planting, recycling of process water, high-efficiency pump systems with variable frequency drives (VFDs), agricultural reuse of lime softening residuals, geothermal heating and cooling, use of regional and recycled materials, and recycling of construction waste.

In addition to ideas for improving the design of new facilities, existing practices were "discovered." Water loss control, reinvestment in infrastructure, preventative maintenance, and water rates that reflect the full cost of service are just a few of the existing practices recognized as sustainable. In this way, the effort helped show the ways in which employees across the organization are connected to sustainability.

LEED certification is not applicable to all projects. Because the program provided a structured approach and goals, it was a useful vehicle for exploring ways to reduce the environmental impact of American Water's facilities. However, the design and construction practices that were identified are widely applicable, good for the environment, and make economic sense for projects of all sizes.

- Optimized pump systems—carefully select pumps for the range of anticipated loads (don't oversize), consider VFDs where beneficial, and utilize SCADA systems to optimize operation.
- Recycled water—where permitted, incorporate properly designed recycle systems to reduce waste of natural resources.
- Concrete—utilize fly ash or slag substitutes if locally available, equal strength, higher reflectivity, recycled material, reduces significant indirect CO_2 emissions from cement production.
- Native landscaping—reduce irrigation requirements, establish prairie or other native vegetation, can reduce maintenance costs (mowing), opportunity to collaborate with local conservation groups.
- Energy-efficient building materials—reflective materials reduce heat gain, conserve energy.
- Lighting and climate controls—conserve energy.

Because water utilities deal with limited natural resources, long-lived infrastructure, and public health, thinking in terms of sustainability is a natural fit.

Contributed by Daniel Haddock, Layne Christensen Company

56. Use Your Building as an Example

Having a LEED certified or Living Building Challenge building is great for your organization and employees. But if that's all you see, you're missing a big opportunity. The building, its operation, its products and processes can be teaching tools, instructing and inspiring others to explore the possibility of sustainable buildings. Give tours often and include signage around your building and its grounds to make sure no teaching opportunities are missed. One of the best things we can do is share our successes and challenges to help clear the way for others.

SUSTAINABLE PURCHASING

57. Use Life-cycle Costing

Life-cycle costing (LCC) takes into account the initial cost of an item, its maintenance costs, and its disposal cost. It is useful for everything from vehicles to buildings. The operating, maintenance, or labor costs associated with a purchase can often be well more than the item's initial or "first" cost over the life of the purchase. See appendix L for a more complete explanation.

Yarra Valley Water in Victoria, Australia, began actively working toward sustainability in 2002 and has embedded it ever more completely into their culture and decision making with each passing year. Appendix M presents the work that they undertook to help their staff select more sustainable infrastructure options. It uses triple bottom line thinking, life-cycle analysis, life-cycle costing, and total cost assessment methodology. This is a great example of how your utility's ideas can be broadened by looking at what you do through the lens of sustainability.

58. Use 100% Postconsumer Waste Recycled Paper

Bill stock, copy paper, letterhead, brochures, forms—you can print almost anything on paper that contains postconsumer waste. That is paper that has been used by others and sent through a recycling system back to the manufacturers.

It may take some extra testing with bill processing equipment or outside vendors if you send your bills out through a mailing house, but 100% PCW paper should not automatically be disqualified because others think it is too dusty, too flimsy, or just not up to the job. This paper is made to the same standards as other paper, it's just processed differently.

Think it doesn't make a difference? By using 100% PCW paper instead of 30% PCW for its 600,000 bills, TVWD saves more than 150 trees, 64,000 gal of water, 7,000 lb of solid waste, 100 million BTUs of energy, and 14,000 lb of greenhouse gases each year. Another fact: it takes about 13 oz. of water to make one sheet of nonrecycled 8.5 × 11 copy paper. So it benefits water utilities in two ways.

TIP: Remember that recycled paper has "bad batches," just like nonrecycled paper. If some paper doesn't work quite like you want it to, try it again. Don't assume that all recycled paper is unsuitable for your purposes.

ANOTHER TIP: When you begin to use recycled copy paper (30%, 50%, or 100% PCW), don't make a major production out of it. Simply start using it. If you say you're using new paper and want to hear if there are any problems, then you'll get complaints just because some people are looking for every little thing to be an issue. Better to just do it, and if there are any real difficulties, you will no doubt be informed.

59. Purchase Office Products Made From Recycled or Otherwise More Sustainable Material

Create a hierarchy for purchases such as the following:

a. Items no longer needed by one employee are returned to the supply room for use by another. From file folders to staplers, items are reused before new ones are purchased.
b. Supplies made from easily recycled materials such as cardboard, metal, or glass are preferred to plastic. Recycled content is preferred, even in easily recycled materials.
c. When no items with easily recyclable materials are available, plastic items containing high recycled content are purchased.
d. Only when no items meeting the above criteria are available are virgin plastic items purchased.

60. Request Office Supplies Be Delivered in Reusable Bins

If your suppliers will do this, choose to receive small items, such as office products, in reusable bins instead of the usual less-reusable cardboard boxes with packing material. This may save the supplier time and materials, as well.

61. Minimize Deliveries

If you don't need paper clips today and pens tomorrow, wait to order until you have a reasonable size order or need something right away. This may involve staff training if personnel are used to getting something immediately upon request. Add a step to the ordering process that requires a date needed on the product so it isn't ordered prematurely.

62. Purchase Duplexing Copiers

It is rare now to find a large copier that does not print double sided, but be sure any that you purchase have that capability. The more challenging part is training users on this feature. If the equipment also handles printing from staff computers, then the best idea is to program it to default to double-sided copies. If a single-sided document is needed, that is the setting that must be consciously chosen.

63. Minimize Purchase of Desktop Printers

You probably have more people with desktop printers than truly need them. Reserve these only for employees dealing with sensitive information. Set default printing to central printers with duplexing capabilities. This will reduce expenses, electricity usage, the material used to make that many more printers and accessories, and likely time and effort in maintenance and stocking of cartridges, etc.

64. Minimize Purchase of Personal Heaters

Personal heaters wreak havoc on the ability to set the temperature in most buildings. In addition, most heaters draw a lot of energy. Some facilities managers have just said the heat is going to be constant, and personal heaters are not allowed. In some settings, wearing an extra sweater is the answer. If someone must have a heater, investigate ceramic heaters or electric panel floor heaters, as they are more energy efficient. Make sure employees are not bringing in their own unless facilities staff has checked the safety and energy efficiency of the unit. See the Resources section for an article on various types of heaters.

65. Purchase Recycled Janitorial Products

Most janitorial paper products—tissue, toilet paper, paper towels, etc.—are available with high recycled content. The percentage available does vary with the product. Toilet paper and paper towels are easily available in 100% PCW (postconsumer waste), while facial tissue's content is much lower. Consider, also, whether you really need the products to be bleached white. Kraft-colored items generally require fewer chemicals in their manufacture, but again, check the particular products with your vendor.

66. Use Rechargeable Batteries

If performance is adequate, use rechargeable batteries in digital cameras, cell phones, and other equipment as appropriate.

67. Purchase Energy Star, Greenpeace-rated, or EPEAT-rated Computers and Equipment

Most electronic equipment is Energy Star rated now, so that should be easy to find. To go the next step, investigate the Electronic Product Environmental Assessment Tool (EPEAT) rankings or the Greenpeace International "Guide to Greener Electronics" Web sites listed in the Resources section of this chapter. Some of these may cost a bit more (though don't just assume that), but you may be getting higher quality or less electricity usage with some of these models, which could mean longer life and a lower life-cycle cost. Remember to check out all the attributes, operations and maintenance cost, and disposal cost before making a purchasing decision. See appendix L on life-cycle costing for more details.

EPEAT rankings are based on environmental criteria, and computer manufacturers register each individual computer model. The gold, silver, and bronze rankings provide an easy classification to understand. In contrast, Greenpeace compares electronics manufacturers according to their general practices on toxic chemicals, recycling, product life cycles, and organizational sustainability. Individual computer models can have a high EPEAT score while the manufacturing practices as a whole could be ranked lower by Greenpeace. So for bulk purchasing or for setting an organizational standard, it is best to review both guides.

68. Purchase WaterSense-Rated Items

Toilets, faucets, showerheads, and landscape irrigation services are all eligible for USEPA WaterSense certification. When you purchase these items, make sure they have the WaterSense label. In order to be considered for this label, products must be tested and certified to meet the criteria in the WaterSense specification for water efficiency and performance.

This generally means these products will:

- perform as well or better than their less-efficient counterparts,
- be about 20% more water efficient than average products in that category,
- realize water savings on a national level,
- provide measurable water savings results, and
- achieve water efficiency through several technology options.

The WaterSense label also appears on professional certification programs for landscape irrigation professionals. These WaterSense-labeled programs verify professional proficiency in water-efficient irrigation system design, installation, maintenance, and auditing.

69. Purchase Reusable Plates/Glasses/Silverware Instead of Disposables

Reusable items such as plates, cups, glasses, flatware, and refillable writing utensils are usually preferred to disposables. And don't feel the need to stock everything for everyone. You may be able to stock just a few types of pens, all refillable, instead of two dozen of every shape and kind. It's good to make employees happy, but it's also okay to lay down a few rules in the name of sustainability!

70. If Disposables Are Needed, Purchase Biodegradable Items

Investigate using disposable plates made from substances such as cassava, sugarcane, limestone, potato starch, corn starch, and recycled fiber. Before you buy, however, check with your recycler or composter to see how these substances will actually break down in the real world. Be careful of biodegradable plastics, for instance; they must go into the proper environment in order to break down well. Otherwise they are just trash.

71. Remove Bottled Water From Facilities

This is not only a testimonial to the value of your product but reduces the resources needed to obtain and bottle the water, make the plastic bottles, and transport the product. It also reduces the waste from the used bottles (only about 12% of plastic bottles are recycled) (MSNBC, 2009).

In 2006–2007, Barwon Water (Australia) received its first Waste Wise Certification for their Ryrie and Lonsdale Street offices. The certification was renewed in 2008–2009. The Waste Wise program is a partnership between Sustainability Victoria and the Barwon Regional Waste Management Group.

The program recognizes the development of a waste-reduction action plan for the organization. Actions completed have included:

- A system for recycling batteries and fluorescent tubes, and trialing composting organics.
- Recycling of old showerheads.
- Setting up a user-friendly system for employees to take trains to Melbourne.
- A Sustainability Victoria energy audit.
- Installation of data loggers on Ryrie Street (head office) and South Geelong offices linking to the Web and providing real-time water use data. Has an alert system to advise of greater water use within set limits.
- Distribution of reusable bags to all staff.
- 35% reduction in paper usage.
- Introduction of the "Golden Palm" staff competition, awarding points for out-of-work activities to encourage more sustainable living practices.

Proposed Waste Wise actions include:

- expanding a trial for staff to take home organic waste;
- implementing a staff eco calculator;
- introducing a computer "switch off" message to staff at the end of each day;
- undertaking ongoing waste audits every three months to monitor change—aiming to reduce the number of waste skip collectors per week by reducing waste and encouraging recycling;
- assisting procurement with introduction of a sustainable purchasing policy;
- engaging with companies that are like-minded/showing environmental leadership;
- reporting "Green Team" progress stories in the weekly staff newsletter, *Connections,* and including regular sustainability tips for staff;
- introducing a regular garden produce exchange, for staff to bring in excess fruit and veggies to exchange for other produce; and
- participating in Clean Up Australia Day.

The Waste Wise certification follows previous work including:

- recycling ink and photocopier cartridges, computers, monitors, photocopiers, mobile phones, showerheads, milk cartons, glass bottles, magazines, cans, and plastics;
- a 15% reduction in paper usage since 2002; a further 30% is expected through an advanced electronic document management system;
- using environmentally sensitive cleaning chemicals;
- trial of waterless urinals;
- air-cooled air-conditioning; and
- a computer switch-off policy (Barwon Water, 2009).

72. Serve "Family-style" Meeting Meals Instead of Individual Servings

You just have to ask many of your favorite takeout places what they can do to cut down on what you have to throw away when you buy from them. Don't accept 20 clamshell boxes with bagel sandwiches, chips, and a cookie, when a selection of sandwiches could be put on a returnable tray next to a tray of cookies and a bowl of chips (either individually bagged for sanitary reasons or in bulk).

Sandwich ingredients could also be laid out so participants can make their own sandwiches. We know of one vendor providing this type of layout who put condiments in hollowed-out green peppers instead of the usual disposable bowls. These vegetable "bowls" can then be rinsed out and composted on-site if not eaten. Caterers or sandwich shops may not think to do this on their own, but help them figure out what will make you happy and save them money on plastic.

Create a list of approved vendors who are willing to work with you to reduce waste, serve local or organic food—whatever your priorities are. Let your staff know that those are the places they can order from. Send a letter or give a call to those not on the list to tell them that whenever they are able to offer more sustainable options, you'd like to work with them.

Ask for canned or glass bottled drinks, not plastic, whenever possible. As a water utility, you likely do not purchase bottled water for staff or on-site events, so just invest in a few good glass pitchers to promote the healthy tap water you serve!

WASTE REDUCTION/ POLLUTION PREVENTION

73. Conduct Waste Audits

Periodically conduct waste audits to determine recycling effectiveness. See chapter 3 for discussion. And look at your organization as a whole. Where do you see waste and what can be done about it? Programs may exist to reward your good works, as well (see p. 83).

74. Use Oil/Water Separators

An oil/water separator keeps oily runoff from wash racks and fueling stations from reaching the stormwater system.

75. Burn Waste Oil or Collect and Return for Rerefining

Collect old oil and return to a facility for rerefining whenever possible. Burning waste oil to heat a shop is not ideal, as it creates carbon emissions, but is better than sending it to a hazardous waste facility that disposes of it.

76. Use Laptops, Scanners, and PDAs as Appropriate

Enter field information, including mapping, repair, and maintenance logs, on laptops and upload the information to your utility's main computer system, reducing the need for additional paper and duplicate labor.

The Medford (Ore.) Water Commission recently completed a project to improve the dewatering and drying of water treatment plant residuals. This process uses engineered concrete backwash water lagoons to efficiently produce cake-like dried solids through a combination of gravity thickening and solar drying. The lagoon design, along with carefully managed sludge loading cycles, takes maximum advantage of the regional climate and its characteristic high evaporative potential (see Figures 4.3, 4.4, and 4.5).

The final dried solids are deposited at the local landfill where the residuals are used as daily cover. Disposal fees are significantly reduced due to the nature of the material. Medford has implemented a simple, energy-efficient process to produce a residual with very low moisture content that minimizes transportation and disposal costs while providing additional benefits for landfill operations—a winning combination for sustainability.

Figure 4.3. Typical backwash sludge after decant

Figure 4.4. Residuals from 600 mil gal production; left is after 50 days, right is after 33 days

Figure 4.5. Dried residuals ready for landfill

77. Use Natural Soil-retention Products

Use bark bags, straw wattles, sediment fences, and other products to increase soil retention, decrease erosion, and reduce silt near storm drains. Install silt fences around debris piles, and move spoils to less vulnerable areas.

78. Make Water Diversion Equipment From Discarded Bike Shop Inner Tubes and Sand

Sometimes the most effective tools are created from what has previously been considered waste. A Tualatin Valley Water District valve flushing crew found a bicycle repair shop with bike inner tubes they usually trashed. The staff took several dozen, filled them with sand, and tied the ends to make sandbags for diverting water when flushing hydrants. The practice continues and is beneficial to both the water utility and the bicycle shop.

79. Use Safer Water Treatment Chemicals

This is another area where there is no one best chemical, but it is important to consciously decide what to use. Optimize your chemical use so no more is used than needed. Maximize the use of water sources or plants that don't need as much treatment. Consider where the chemicals are manufactured, i.e., buy from as local a vendor as possible to minimize transportation costs and emissions. Think about disposal, too. Hire an environmental waste disposal company if needed to dispose of chemicals responsibly.

80. Handle and Dispose of Residuals Carefully

Give careful consideration to the way you handle residuals. Are these solids of a composition that you can land apply to fields? How far do you have to haul them for disposal? Do you have to haul them wet, which is heavy and results in greater fuel use and more resulting carbon emissions? Think through the process and options before you landfill.

81. Return All Unusable Metal (Cast Iron, Brass, Steel) to the Utility for Sale as Scrap

Gone are the days of throwing scrap in trenches. Not only is it environmentally undesirable, but there is money to be made from almost any metal a utility uses. It is worth bringing it back to the shop, storing it, and finding someone to buy it.

82. Use Dilute Hypochlorine Solution Instead of Gaseous Chlorine at Pump Stations

The hypochlorine solution is safer for the environment and employees and is as effective as the more hazardous gaseous chlorine.

83. Utilize Directional Boring for Laying Pipe Instead of Trenching

Directional boring is appropriate for some projects. If you have the equipment or can rent it, directional boring for pipe installation will reduce the disruption to above-ground areas. This means less restoration needed, with associated reduced cost and improved community relations in some cases.

84. Cover Spoils From Construction With Straw, Hay, or Jute Matting

As your conservation specialists will tell you, this will hold in moisture and reduce erosion while the area you have disturbed returns to its restored state.

85. Minimize Vegetation Removal Required at Construction Sites

86. Use Lead-free and Toluene-free Paints for Locating Services

Read the labels. Most spray paints are available in these more environmentally sound varieties.

87. Separate and Recycle Concrete and Asphalt Pieces Cut Out of Streets for Leaks and Repairs

It is surprising how much money you can save by not taking these heavy materials to the dump. If at all possible, separate these materials and find out if they can be taken to a separate area where they can be reused and you won't have to pay to deposit them.

88. Reuse Recycled Rock and Ground Tailings in Construction Projects Whenever Appropriate

89. Practice Unidirectional Flushing

Instead of just opening the hydrant and allowing water to flow until it appears clear, practice unidirectional flushing. Starting from a clean source, isolate all sideline valves and the nearest mainline valve beyond the discharge hydrant, creating a one-way flow. This will allow you to systematically flush from the "dirty" to "clean" sections of the main. Unidirectional flushing is a more efficient way to remove sediment and biofilm within pipes and reduces the amount of water used in the process.

90. Reuse Any Functional Parts From Old Meters and Hydrants to Repair Other Meters and Hydrants

If you have room to store old meters and hydrants for a while, this is a good way to save money and reduce waste. When no usable parts are left, sell any remaining metal as scrap.

91. Use Reusable Aluminum and Wood Signs at Construction and Maintenance Sites

92. Partner With Local Parks and Recreation Agency

Use vacant utility sites to dump landscape waste from both agencies. You don't have to pay to dispose of landscape waste and the parks and recreation folks get feedstock to grind and mulch for trails, etc.

93. Use Fewer Toxic Materials

Develop a list of toxic materials used in your operations and work to replace these products with less harmful substances whenever possible. Use an existing inventory list or take the time to list what products you use in your operations, from fluoride to light bulbs to cleaning supplies. Your safety coordinator may be able to provide a list of products requiring MSDS sheets. That may be a good place to start, as those products are likely the most toxic materials used in your operations. Ask the appropriate employees to look for alternatives to those with the most toxic ingredients and try them. If they do the job, consider switching to these alternatives for both sustainability and employee safety reasons.

94. Promote Online Phone Books

And refuse delivery of unneeded paper copies. Most phone companies offer online (Web- or CD-based) phone books. Make sure your front desk staff knows to refuse other paper phone books that may be automatically delivered. If left outside your door, don't just recycle them. Call to ask that they be picked up and you are less likely to have them delivered again next year. If you do want a few paper copies for certain areas, call your preferred vendor and ask for the number you want.

95. Request Sustainable Practices of Presenters

Ask anyone who comes to present at your utility to bring their presentation in electronic form (on a thumb drive or CD) so you can keep a copy that employees can access. In most cases, it is not necessary for presenters to bring handouts for everyone. If presenters feel it is important to bring handouts, then request double-sided copying and, if you have 100 employees, maybe only 75 copies are really necessary. At the presentation, encourage employees to take the handouts if they have a use for them, but leave them at the door if they will not use them. This may allow the presenter to use the paper for another group, or at least allow the paper to be reused or recycled.

96. Use Printed-one-side Paper for Draft Copies or Make Into Scratch Pads

Encourage duplex copying as the norm. Set the default on printers and copiers to double-sided copying whenever possible, so employees must make a conscious choice to make a single-sided copy.

If there is a reason for one-sided printing, reuse the printed-one-side paper for drafts. Paper that has come through a color copier that uses wax-based ink cannot be rerun for drafts as the color runs when put through the heat of another printer run. Collect that paper in marked boxes in copy rooms. Printing companies are usually happy to cut that printed-one-side stock in halves or quarters, back the paper with chipboard, and make them into pads for little or no cost.

97. Distribute Large Documents Electronically

Do you routinely distribute large documents, such as Board agenda packets or requests for proposals (RFPs), to numerous people at once? Consider providing Board meeting packets on thumb drives or CDs to reduce paper usage. Make RFPs available online or, if you want to know who receives one, make it available electronically by contacting the contracts manager. You will save paper, ink, postage, the time of making copies, as well as the carbon emissions associated with all of those endeavors.

98. Encourage Employee Recycling

Place recycling bins at every desk, in common areas, and by copiers and printers to encourage recycling of paper and other items. Often, all recyclable items (except glass and plastic bags) can be placed in desk-side bins for convenience.

99. Use a Rubber Stamp for Incoming Payables Instead of Attaching a Paper Form

100. Put Employee and Customer Forms Online

Put as many employee and customer forms as possible online—print only as needed.

Remember a few trainings and reminders will likely be needed to get employees used to looking online instead of on a shelf.

101. Implement a Paperless Plan Review Process

While less paper is used for construction plans and proposals, you can do even better by encouraging, then requiring, that all documents in the construction process, from submittals to shop drawings, be handled electronically. The same goes for developer and design proposals.

102. Develop an Automatic Electronic Archiving System

Decrease paper reports by allowing them to be viewed without printing. This, of course, requires some IT expertise, but if you have it on staff or can contract it out, an electronic archiving system can save a lot of paper, staff time spent looking through reports, and the storage space for those reports.

Technology has made paper use far less necessary for all of us. Encourage electronic communication when documents need to be transmitted. Make sure scanning and archiving is possible from your main copiers so items not previously in electronic form may be saved electronically.

> TIP: Be sure to research which software is the best for your plan review process. All parties involved need to agree on the same products to streamline the process. Adobe Acrobat® is emerging as a likely choice at present.

TIP: When you develop your electronic archiving system, be sure to establish and manage a cataloging system for electronic references as you would for physical documents. Files dumped in unrecognizable places will do you no good, but if you develop a recognizable repository, much time, space, and paper may be saved.

Look for electronic archiving opportunities such as:

- finance and accounting monthly reports, budgets, and comprehensive annual financial reports (CAFRs); and
- engineering capital improvement plan (CIP) reports, CIP tracking modules, and progress reporting.

Develop an electronic reference library with items such as:

- AWWA Standards;
- publications;
- administrative policies, resolutions, manuals;
- maintenance manuals;
- lab procedure manuals; and
- vendor catalogs.

103. Put a Line at End of E-mails Discouraging Printing

Include a line after your signature like: "Please consider the environment before printing e-mails" or "Please only print e-mails if necessary."

104. Route Magazines and Documents to Interested Parties Instead of Ordering or Printing Copies for All Readers

105. Foster Innovation

Challenge, promote, encourage, and recognize employees who use trash to make new items, suggest waste-reduction actions, find new uses for printed-one-side paper, etc.

106. Compost Coffee Grounds, Fruit and Vegetable Scraps, Tea Bags, etc.

If your community has commercial composting, participate in that. If not, a simple compost bin or two, with a few staff people (sustainability team members, perhaps) assigned to take out and turn the compost could do the trick. Find places you can use the finished compost, or partner with a local community garden to take it. If you have the room and employees are interested, develop an employee garden at one of your sites and donate harvested produce to a food bank.

107. Use Nontoxic Cleaning Supplies

Use products that carry the Sustainable Earth, Green Seal, or USEPA Design for the Environment labels. You can probably reduce cleaning supplies to four or five formulations that come in bulk containers and are dispensed from a cabinet.

Waste causes great loss of value and resources. Humans are the only species that create waste. We can learn to identify all types of waste and, through their elimination, save money and achieve a more sustainable world.

The visionary goal of Zero Waste expresses the need for a closed-loop industrial/societal system as suggested in Figure 4.6. Waste is a sign of inefficiency. Our use of the term ***Zero Waste*** includes zero solid waste, zero hazardous waste, zero toxics, and zero emissions.

Zero Waste, a visionary goal that strives for:

- **Zero Waste of resources**
  –100% Efficiency of energy, materials and human resources
- **Zero Solid Waste**
- **Zero Hazardous Waste**
- **Zero Emissions—to air, water, or soil**
- **Zero Waste in Production Activities**
- **Zero Waste in Administrative Activities**
- **Zero Waste in Product Life Cycle**
- **Zero Toxics**

Resulting in:

- **Reduced risks to employees**
- **Reduced risks to the environment**
- **Reduced presence of toxics creates less hazardous waste**
- **Closed loops for materials**
- **Reduced costs**

Figure 4.6. Goals of Zero Waste

Zero Waste suggests that the entire concept of waste should be eliminated. Instead, waste should be thought of as a "residual product" or simply a "potential resource" to counter our basic acceptance of waste as a normal course of events. Opportunities such as reduced costs, increased profits, and reduced environmental impacts are found when returning these "residual products" or "resources" as food to either natural or industrial systems. This may involve redesigning both products and processes in order to eliminate hazardous properties that make them unusable and unmanageable in quantities that overburden both industry and the environment.

Zero Waste strategies consider the entire life cycle of our products, processes, and systems in the context of a comprehensive systems understanding of our interactions with nature and search for inefficiencies at all stages. With this understanding, wastes can be prevented through designs based on full life-cycle thinking. Indeed, we should work to "design" our wastes, if any, so that they have future applications.

The comprehensive nature of a Zero Waste strategy is shown in the following input-output diagram:

(continued)

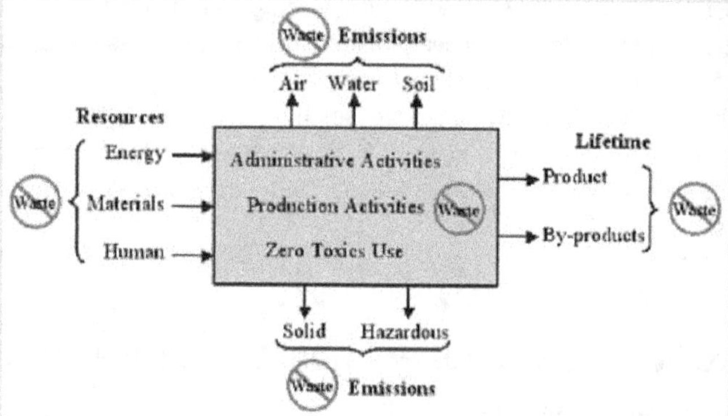

Figure 4.7. Application of Zero Waste strategy

A Zero Waste strategy leads us to look for inefficiencies in the use of materials, energy, and human resources. To achieve a sustainable future, extreme efficiency in the use of all resources will be required in order to meet the needs of all of the earth's inhabitants. See www.zerowaste.org for more information (ZWA, 2009).

Your cleaning staff then adds water to achieve the proper dilution and is able to clean everything from glass to floors. This move will reduce the cost of purchasing small amounts of many products, reduce the packaging used for those smaller products, and simplify work for the custodial staff. This would not be a good move if cleaning results suffer, so test all products to make sure they work well before a transition is made.

Reduce, Reuse, Recycle

Reducing your usage of products and the packaging that comes with them is ideal. But the reality is that we use materials when we provide water to our customers. So our challenge is to do that while producing as little waste as possible. The Zero Waste Alliance would actually have us rethink the model where anything is even considered waste. See the sidebar on p. 91 for ways you may be able to rethink your process and come up with modifications that reduce or eliminate what you now consider waste.

108. Increase Recycling Awareness With Education and Signage

Some employees respond better to text, others to visuals. Experiment with both on your recycling signage to see what resonates best. In addition to signage at the point of recycling, include tips, facts, and reminders in employee newsletters or signs posted in common areas. Revolving newsletters in bathroom stalls may get read by the captive audience!

Periodically, make presentations at staff or crew meetings. A virtual tour (PowerPoint slide show) of the site's recycling opportunities is a good reminder of what is recyclable and where to take it. A game show format PowerPoint presentation asking employees whether or not certain items are recyclable is fun (especially if you throw small prizes to those answering correctly), interactive, and a good way to update coworkers regarding what can be recycled or reused as this information changes.

> *Zero Waste is a whole-system approach to redesigning the flow of resources through society.*
>
> *Warren Snow,*
> *Zero Waste New Zealand*

109. Create an Employee Recycling Center

Create an employee recycling center to allow staff to bring in items they are unable to recycle at home. Educate custodial staff regarding placement of collected materials in appropriate waste-hauler's containers. Place well-marked containers for authorized materials and food in lunch and break rooms. Make sure every desk and common area has a recycling bin. Where possible, put the recycling bin in front of the garbage so it is easier to use.

110. Send Everything Possible Out for Reuse or Recycling

If you have materials that are going to the landfill, know that many items can be reused in some way, and many more can be recycled back into new products. The following is a list of those items reusable or recyclable in the Portland, Ore., area (items 111–136). Items may be different in your area, but ask your hauler if they will take them. Tell them there are areas of the United States where these items can be reused or recycled. Asking the question is important, even if you get a negative answer today. If no one asks, it is certain to be much longer before you have options like this in your area.

111. Alkaline Batteries (from pagers, locaters, cameras, cell phones, etc.)

112. Block Styrofoam

113. CDs/DVDs

114. Cell Phones

These are collected by nonprofits, military programs, schools, and some manufacturers, then refurbished and given to victims of domestic violence and other beneficiaries.

115. Construction Debris

Ensure that everything possible from construction or maintenance activities is reused or recycled. Asphalt, metals of all kinds, wood, and more can be kept out of the landfill.

An organization called Free Geek (www.freegeek.org) that started in Portland, Ore., is an example of a creative way to reuse electronics before sending them to be recycled. Their tag line is: "Helping the needy get nerdy since the beginning of the third millennium."

Their mission is:

To recycle technology
and provide access to computers, the internet,
education, and job skills
in exchange for community service.

Since February 2000, they have taken arguably obsolete electronic materials from all over the Portland area and beyond. Believing firmly that reuse is the highest and best form of recycling, they refurbish as much of that material as they can and send it back out into the community. They do their very best to make sure that all electronic materials they don't reuse are responsibly recycled within North America. Free Geek recycles approximately 500 tons of scrap per year and puts thousands of refurbished computers back into the hands of people who will use them.

There are now others who have started similar services across the United States and in Canada. See if there is anything similar in your area.

Free Geek Portland (Portland, Ore.)
Free Geek Arkansas (Fayetteville, Ark.)
Free Geek Central Florida (Orlando, Fla.)
Free Geek Chicago (Chicago, Ill.)
Free Geek Columbus (Columbus, Ohio)
Free Geek Michiana (South Bend, Ind., area)
Free Geek Twin Cities (Minneapolis-Saint Paul, Minn.)
Born Again Technologies (Murfreesboro, Tenn.)
Free Geek Vancouver (Vancouver, B.C.)
Free Geek Providence (Providence, R.I.)
(Free Geek, 2009)

If you are actually tearing down a structure, consider the benefits of "deconstruction" and salvage services that remove all reusable items before the rest is demolished.

116. Electronic Items

Before you even buy these items, ask what the manufacturer's take-back program is. Manufacturers are beginning to build these items for disassembly and reuse. Use those manufacturers that are in the forefront of those efforts.

Your next step is to see if there are any schools or nonprofits that might be able to use these items.

If you do need to recycle electronic items, do your homework! Do as much as you can to make sure your collector does not ship these items out of the country for disassembly in unsafe conditions. There will likely be a fee for collecting these, because there is a lot of work that goes into separating what can be salvaged and disposing of the rest responsibly.

117. Floppy Disks

Schools might have use for these, but they are rapidly becoming a thing of the past. See if whoever takes your CDs and DVDs will take these also.

118. Glass

This needs to be collected separately so broken glass does not contaminate other recyclables.

119. Lawnmowers

These can be refurbished and resold if you can find a retailer or individual willing to do this.

120. Light and Corrugated Cardboard

121. Light Bulbs

Incandescents, compact fluorescents, fluorescent tubes—most light bulbs are recyclable, but there may be a fee.

122. Metal (copper, aluminum, iron, tin)

You will probably be getting paid for these materials, so there is no excuse for burying these in the ground or throwing them away!

123. Meter Registers

If you install temporary registers at construction sites (in case they get damaged), and install the more expensive ones once homes are ready for occupancy, you may end up with lots of temporary registers and no place to put them but in the trash. Ask your local recycler if they are willing to disassemble them for the metal and separate the glass, plastic, etc., for recycling, as well. A recycling facility in Hillsboro, Ore., does that, which saves many pounds of registers from going in the trash and provides a small income stream, as well.

124. Motor Oil

Send motor oil and antifreeze back to facilities that will rerefine it and sell it again. Good used motor oil might also be burned in a shop furnace for heat. This is not a great solution, but it lies somewhere between returning the oil to be rerefined and throwing it in the garbage or dumping it down a drain, which is illegal.

125. Newspaper

126. Ni-cad Batteries

127. Packing Air Bags and Packing Peanuts

Most shipping stores are happy to get these now so they can reuse them and not have to buy new. Just take them to a UPS store; Mailboxes, Etc.; or any of the multitude of smaller businesses that offer shipping services. Everybody wins.

128. Paper

129. Plastic Bags

These get caught in recycling equipment if thrown in with the other recyclables, so be sure to collect these separately and check with your hauler to see if they are even able to accept them.

130. Plastic Bottles and Other Plastics Accepted by Your Hauler

131. Printer/Toner Cartridges

132. Telephone Books

Remember to opt for the online version and consciously reduce the number of paper phone books you get and how many publishers you get them from.

133. Tennis Shoes

Nike takes old tennis shoes as part of their Reuse-A-Shoe program started in 1993. The purpose of the program is to eliminate waste and close the loop on Nike's product life cycle by collecting postconsumer, non-metal-containing athletic shoes of any brand.

Once collected, the sneakers are ground up and separated into three distinct types of material: rubber from the outsole, foam from the midsole, and fabric from the shoe's upper. Nike and its partners take the granulated rubber and create soccer, football, and baseball fields; weight-room flooring; and running tracks. Nike uses the granulated foam from the shoes for synthetic basketball courts, tennis courts, and playground surfacing tiles. The granulated fabric from the shoe uppers becomes the padding under hardwood basketball floors.

These programs are located in Australia, Canada, Japan, the United Kingdom, and the United States. See http://www.recycledrunners.com/ for drop-off locations.

134. Uniforms

This is a positive for safety and security as well as sustainability. If you can find a reputable business that cuts up old materials for rags, they will cut out the uniform logos, making it impossible for the uniforms to be used improperly. Educate employees from a security angle, requesting their cooperation or requiring old uniforms be turned in before new ones are issued.

135. Usable Shoes

Send usable shoes to organizations such as http://www.soles4souls.org. This organization donates usable shoes to those who need them all over the world. Soles4Souls has ways for you to get donated shoes to them from anywhere.

136. Wood Pallets

Ask if your vendors will take these back and reuse them, or find others who can use them before you recycle this item.

I hope there are even more items that you can reuse or recycle where you are. The less that goes in your dumpsters, the less you have to pay for garbage service and the slower landfills reach the end of their lives. Be creative and resourceful. You might even get some money or a boost to your community image if you find a unique way to use what you thought was waste!

Transportation Issues

(employee commute trips, customer trips, company fleet operations)

137. Implement Leadership in Energy and Environmental Design Features such as Carpool Spaces, Bike Storage, Employee Showers

138. Trip Planning

Plot routes in advance to reduce unnecessary mileage on fleet vehicles. From meter reading routes to combined errands, a little coordination will go a long way toward reducing excess trips, fuel usage, and carbon emissions.

139. Establish Satellite Offices or Check-in Sites

If employees reporting to construction sites or reading meters in a location far from your headquarters can save time and fuel by traveling directly to their work site, explore ways to make that feasible. You may need to establish policies regarding allowing personal vehicles to be driven to work sites or utility vehicles to be driven home. GPS may allow you to establish the location of work vehicles, to ascertain timely arrival. This won't work for every utility, but it is a practice to be evaluated.

140. Require Reduced Vehicle Idling in Shop Yard and at Work Sites

Educate employees to promote reduced vehicle idling in your organization's yard. Some vehicles at jobsites may have to be left on in order for emergency flashers to work. New batteries are available for vehicles that may allow flashers/safety lights to work while the vehicle is turned off without compromising the vehicle's ability to start up again. LED beacons are also available. They pull much less energy, so they don't drain the battery as quickly, and again, idling is not necessary to keep lights on and workers safe.

141. Make Proper Vehicle Maintenance a Priority

Create daily and weekly checklists to be completed by drivers to ensure that vehicle tire pressure is appropriate and all maintenance is kept up-to-date. Vehicles that are in proper working order are safer and better for the environment.

142. Evaluate Your Vehicle Purchases

Purchase electric, hybrid, plug-in hybrid, or vehicles with higher mpg and lower emissions than your current fleet. Give pricing preference to purchasing these types of vehicles, or at least give preference to purchasing crew cab vehicles to accommodate four people to eliminate sending two pickups to a jobsite.

143. Use Alternative Fuels—Biodiesel, Ethanol, etc.—as Appropriate

Currently, using some percentage of biodiesel means fewer emissions from the creation and burning of diesel-fueled vehicles. Even bio-based fuels are not perfect (for example, is too much corn being used for fuel and driving up food prices?), and as time goes on, there may be better solutions. Check out the properties of alternative fuels, such as gel point and solvent properties, to make sure they work for your situation.

144. Examine Your Oil Usage Practices

Purchase rerefined motor oil for use in fleet vehicles. As with 100% PCW paper, just start using this. No big "to do" about changing, as that invites people to discover problems. If there are any, you will know.

Consider checking the oil by sample and replacing only when needed, not by miles or date of last oil change. And explore "green" oil filters that may enable you to change your oil less frequently by keeping the oil itself cleaner longer.

145. Use the Appropriate Vehicle for the Job

Evaluate the vehicles assigned to each position or crew and determine whether or not they are the right size for the application. If a locater or field customer service rep is driving around in a four-wheel drive, half-ton pickup, they may be able to use a smaller, more fuel-efficient vehicle. If a crew is taking three two-person pickups to a jobsite, maybe a larger pickup with a crew cab would be more appropriate.

146. Coordinate Errands

Make sure several employees aren't going to the same store the same day. Establish a simple calendar in Outlook® where employees can post the stores or suppliers they are going to that day. If someone else is already going, combine the trips. If you don't have enough employees to warrant an Outlook calendar, just be sure to ask around before you go!

Encourage employees who drive to a deli or store to get a sandwich to handle lunches this way, too.

147. Provide Free or Discounted Transit Passes for All Staff Members

The cost for your agency to purchase bulk transit passes may be far less than individual purchases, thus encouraging public transit usage and providing a valuable employee benefit. Ask your local transit agency what they offer to businesses and government agencies. Allowing a flexible arrival time may make it possible for employees to trust that they can use public transit without being late to work.

148. Establish a Free Ride Home Program

To ensure employees can get home in case of an emergency, work with a local cab company or your fleet manager to provide immediate transportation for employees who have biked, carpooled, or taken public transportation. This removes one more barrier to commuting by alternative transportation.

149. Provide Biking Amenities

Make it easy to bike to work by providing safe, secure facilities for bicyclists to store their bikes and gear while at work. Showers help, too, as does a flexible arrival time.

150. Maintain Alternative Sites for Bill Payment

Any time customers can pay a bill while doing business elsewhere provides convenience for customers and reduces additional time on the road.

151. Provide Online Bill Notification, Bill Payment, and Payroll Options

Many agencies have instituted online billing and payment options for their customers, which continue to reduce the amount of paper needed for financial transactions. Once established, through your own IT department or a third party, be sure to publicize the options to customers frequently, as it may take several times before a particular customer makes the switch.

Additionally, many employees will choose direct deposit for their payroll checks when offered, and many vendors will now accept electronic payments for a majority of your expenses.

152. Provide an Electronic Funds Transfer Payment Method to Reduce Travel and Paper Usage

153. Utilize Teleconferencing

Always evaluate whether or not a face-to-face meeting is needed in any given situation. When possible, encourage conference calls, teleconferences, and webinars instead of meetings involving travel. Some will just involve pushing a few buttons on your desk phone. But more sophisticated technology is also possible. From DimDim® to Webex®, many options are available to provide remotely attended meetings, collaborations, and webinars of your own, in addition to attending others' events.

154. Schedule Meetings to Minimize Driving

Sometimes it is important to meet with others in person. When you do need to schedule a face-to-face meeting with people from varying locations, consider where they will be coming from, if they will be driving or taking another mode of travel, whether or not meetings can be scheduled in conjunction with other events, and what carpool opportunities exist.

155. Provide Incentives for Meeting Attendees to Carpool or Use Public Transportation

One way to encourage sustainable behavior in others is to make it worth their while. Publicize a program to offer a $5 gift card or other small reward to anyone traveling to meetings at your facilities by means other than a single person in a car. Even if you don't carry the program on forever, it will ingrain in some folks the behavior of thinking first about how they travel.

156. Challenge, Encourage, and Recognize Employees Who Use Alternative Transportation

Encourage carpooling, walking, and riding bikes from both sustainability and wellness perspectives. Post a sign-up sheet online and in lunchrooms for employees to find commuting partners. Develop an incentive program for riding bikes to work. Explore the feasibility of using a company van as a commuter van from outlying areas—some rental companies have programs to make that more cost-effective.

157. Allow Flexible Work Hours

Allow flextime, four tens or 9/80 work schedule, or early start to keep vehicles off the road, especially at peak times.

158. Share a Fueling Station With Another Agency and Agree to Some Percentage of Biofuels

159. Participate in Community Events to Cut Down on Single-Person Commuting

Other agencies in your community may have events you can promote at your utility. This reduces the effort of creating your own event and lends support to others working toward the same goals.

Outreach and Education

160. Provide Water Conservation Training to Customers

Assessments, leak detection services, and other training encourages individuals and businesses to use water efficiently through education.

Every July, throughout the Portland, Ore., metro region, companies participate in an event that offers a supportive, fun environment for trying sustainable transportation: the Carefree Commuter Challenge (CCC). Employees track their sustainable commute trips during the month to win great prizes and pats on the back. Transit, walking, bicycling, carpooling, and vanpooling all count, and so does telecommuting. The CCC also creates a friendly competition among businesses of like size for awards and high praise. Sponsors and in-kind contributors make it all possible by donating cash and prizes.

You may be able to partner with another organization already doing assessments in homes. In Portland, Ore., the Energy Trust of Oregon, an independent nonprofit organization, was already offering home energy assessments to homeowners, so several water districts asked that water assessments be included in the same visit, if the homeowner was interested. The cost is less for each water utility than going out on their own because a representative is already in the home, and it is more convenient for the homeowner because he or she doesn't have to schedule two separate visits.

161. Educate Customers Regarding Water Conservation

Conservation programs dedicated to reducing customer water usage through rebates on water-efficient fixtures and appliances; giveaway items such as low-flow faucet aerators, shower timers, and leak kits; and free advice in the form of educational publications, assessments, and workshops reduce the demand on your system. While in the short term this may reduce your revenue, in the long run it will preserve the resource and perhaps delay your need for new infrastructure and therefore bond sales and increased rates.

Actively participate in community and trade events such as:

- Fairs
- District open house
- School presentations/assemblies
- Home improvement shows
- Landscape shows
- Trade association conferences/meetings
- Vendor fairs

Other outreach methods include:

- Water-efficient demonstration garden
- Water-efficient demonstration restroom (for public use)
- Water conservation information kits and giveaways
- Educational displays
- Web site
- Brochures
- Water-efficiency assessments
- Educational signage, indoor and outdoor

Albuquerque Bernalillo County (N.M.) Water Utility Authority (WUA) uses electricity to pump water from an aquifer, treat river water, blend surface and groundwater, and send this water to customers. WUA uses more than 120,000,000 kW·h (120 megawatts) per year from 120 large electric meters. The majority of electric usage is from June to September.

WUA has a long history of managing its energy through its SCADA technology, utilizing the Water Research Foundation's Energy Water Quality Management System (EWQMS) framework, off-peak pumping, and pump efficiency optimization. WUA recently determined that its water system flexibility provided an opportunity to negotiate options for further energy pricing through an energy demand response program.

Energy demand response includes partnering with a utility's energy provider to alter energy use during periods when energy supplies are constrained. Actions include moving energy consumption to off-peak periods and agreeing to interrupt energy consumption at the request of the energy provider. In exchange for this partnership, a utility is provided lower energy rates or paid incentives.

Last year, WUA signed a contract through a third-party company (EnerNoc) with Public Service of New Mexico (PNM), which is the electric provider for the state. PNM offers a Peak Saver program designed to help large commercial electric customers reduce the amount of energy they use during peak-demand days, typically the hottest days of the year. Because WUA is one of the largest electric users in New Mexico, it was expected that WUA could benefit from demand response and shave power when PNM is in a critical peak period. There is no cost to WUA whether they are asked to reduce usage or not.

After a detailed analysis and review of WUA's power consumption in the summer, both parties agreed that WUA could benefit from putting 15 large electric accounts into a contract of 3 megawatts.

WUA's SCADA collects real-time data from each electric meter every minute. By providing this real-time data to EnerNoc, WUA is able to receive more incentive payments because EnerNoc does not have to put monitoring hardware at each site.

The first two-month contract (July and August) included three interruptions. For reducing their electricity usage as requested, WUA received $68,000 from PNM. WUA has now signed a contract for 6 megawatts of interruption over four months (June through September).

Contributed by Steve Conrad, Simon Fraser University

162. Reduce Resource Demand Through Water Conservation Programs

To encourage water efficiency among residential, business, industrial, and government customers, put as much effort as you can into conservation programs. Each entity will place a different emphasis on conservation programs due to their specific business situations, but in the long run, conservation of resources benefits us all.

Instead of refusing to work with vendors who were not able to provide more sustainable options, TVWD chose to spend time educating their suppliers. In addition to one-on-one discussions with vendors and requests for sustainable offerings, District purchasers invited all their vendors to a luncheon meeting. To broaden the impact for all participants, a dozen other government purchasers from other entities in the area were also invited. The first hour was a presentation and tour of TVWD facilities. The next half hour was a (sustainably catered) lunch and open discussion with vendors regarding the obstacles and opportunities they encounter. The last hour was a discussion just among the purchasers to talk about best practices, issues, and possible collaboration opportunities. The feedback was uniformly positive and all participants seemed to benefit in some way.

163. Adopt a Pricing Structure That Encourages Water Conservation

An increasing block or similar pricing structure encourages water conservation by rewarding customers for using water efficiently.

It is important for utilities to send the same signal through pricing as they do in their conservation programs, that is, use water efficiently. If you are sending conflicting messages between your conservation program and your rates, talk to your finance department.

164. Advocate for Water Supply Obtained From Sustainable Sources

What is sustainable will vary from region to region and country to country. Maximum usage of rainwater is usually sustainable but may lead to storage concerns and the construction of dams and reservoirs that are not entirely sustainable. Groundwater usage should not exceed the rate of recharge and may be a better choice than surface water since it doesn't have to be treated as much. Surface water should generally be taken out and put back into a natural source at the appropriate quality, temperature, and quantity.

There is no one answer, but the key is to think about it from a holistic watershed perspective. Look at the total impact of your supply and maximize your use of sources that involve minimal pumping, chemicals, and energy use.

In addition to your own operations, participate in regional groups that look at water from a holistic perspective. Since water issues don't stop at city or county lines, it is essential to look at issues from a broader watershed perspective.

165. Encourage Adoption of Sustainable Practices Among Customers, Suppliers, and/or Peers

Suppliers and customers may not be as educated about sustainable practices as you are. Use customer newsletters, bill messages, community events, radio or TV advertising, open houses, award ceremonies, and trade shows to mention what you are doing and to ask about their practices. Customer water conservation assessments, for example, might provide the opportunity to discuss other energy-saving practices or products (for example, washing machines and hot water heaters) that could reduce a customer's electric bill in addition to saving water. Look for opportunities to expand your positive impact.

166. Make Sure Incentive Items are Useful and Sustainable

Evaluate your "giveaways" to see if they really benefit your customers or are just trinkets that will likely gather dust. Think about what your audience will actually use, then make sure it is manufactured sustainably if possible.

When you gather prizes for employee events, make them useful and sustainable, as well. Stainless-steel travel mugs, reusable lunch boxes, and reusable grocery or tote bags are examples of items that will get your name out in the world instead of stuffed in a drawer.

167. Train New Employees in Sustainable Practices

Once current employees are trained, it is important to develop a system to educate new employees in the utility's sustainable practices and behaviors. Some organizations have new employee training that includes ethics, safety issues, office practices, and other information. Make sure sustainability is included in the information new staff members receive.

Brochures, articles, or even a video from the Internet or a local sustainability group may be a good start. Train a handful of people at once, or have a one-

Helping customers in financial difficulty remains an important concern for Sydney Water (Australia). Sydney Water provides a range of financial assistance to customers in financial difficulty, either directly or through a network of accredited community welfare agencies. One way Sydney Water does this is by funding payment assistance vouchers.

In 2007–2008, the number of customers who requested bill payment arrangements increased by 5.3%. This was equal to about 5,600 extra payment arrangements and requests for more time to pay bills. Customers who are finding it difficult to pay their bill can contact Sydney Water to make payment arrangements, such as installment plans, that suit their financial circumstances.

Sydney Water has taken part in the national No Interest Loan Scheme, which gives customers interest-free loans for water-efficient washing machines.

It has also developed strong links with community welfare agencies, such as the St. Vincent De Paul Society and a network of agencies that work with people from culturally and linguistically diverse communities. Sydney Water is also currently working to extend its financial support to key community groups in targeted areas for a variety of reasons, including their geographical isolation from services, large household size, and socio-economic need.

Sydney Water gave community agencies greater access, via its Web site, to information about community assistance programs and translated 13 fact sheets into various languages. It also offers a free telephone interpreter service for customers who need information in languages other than English.

Contributed by Judi Hansen, Sydney Water

on-one conversation with each employee, if that is practical, so you can discuss issues face to face. Give them a tour of the facility with stops at your employee recycling area(s), composting sites, duplexing copiers, etc. Other employees will continue the education on the job as situations arise.

168. If You Have a Speakers Bureau, Include Sustainability Experts in Your Offerings

169. Make Presentations Regarding Your Progress to Organizations and Trade Associations

Present your sustainability projects and progress not only to groups in the water industry, but groups such as the American Public Works Association, Government Finance Officers Association, American Society of Civil Engineers, The Natural Step, at local, regional, and national levels. Present at conferences, write newsletter articles, include your sustainable efforts when you participate in events. The more people see what can be done, the more apt they are to adopt more sustainable behaviors themselves. Be a good example!

Community Service and Civic Involvement

170. Create an Emergency Assistance Program for Customers

Different opinions exist regarding the appropriateness of a low-income assistance plan to provide a subsistence level of water to every customer, regardless of ability to pay. Whether or not your organization feels able to provide that, you may want to look at providing a program that is able to give assistance on a case-by-case basis. If you do not have the staff to provide this service, look at partnering with a social service organization that can screen applicants and provide a set amount of funds to those who are in need of temporary help.

171. Build Underground Storage Tanks When Possible to Minimize Environmental Impact and Maintain Natural Look of Area

172. Participate in Neighborhood Litter Cleanup Program or Other Community Betterment Event

173. Allow Employees Company Time to Volunteer

Participate in AWWA committees, Water for People, SMART (Start Making a Reader Today) reading program, or one of the thousands of other volunteer opportunities available. An hour a week may lift some employees' morale and do infinite good in your community. That goodwill may even help your organization later in ways you can't even imagine now.

Sustainability and responding to climate change involve many aspects of urban design, individual behavior, and collective behavior. Utilities need to be a part of any local efforts to address these issues. Utilities, whether public or private, are very much a part of both the problem and the solution.

Douglas Yoder, Miami-Dade Water and Sewer Department

174. Encourage Employee-Supported Donations for Local Nonprofit Groups

Allow employees to adopt a nonprofit organization for which they raise money each year. Your employees will no doubt come up with lots of ways to do this. Some examples: Have employees bring items, new or gently used, to auction off to other employees. Run a food, coat, or blanket collection, with items given to local shelters and food banks. Organize chili cook-offs, dessert bake-offs, or corned beef and cabbage on St. Patrick's Day—just charge a nominal amount to enjoy the fruits of others' labors.

Are there a couple of college sports teams many of your employees support? Run a contest where the team supporters that raise the least amount of money have to wear the other team's colors for a day. Or auction off items with team logos to raise the most money. Anything you can do to give employees a little value, and give them a chance to contribute to the greater good, will be appreciated by all.

175. Participate in Conservation Programs, Educational Fairs, and Other Community Events as Appropriate

176. Take Into Account Minority/Women-Owned and Emerging Small Business Status in Purchasing

177. Produce a Green Team Resource Book Available for Employee Use

178. Share Unused Property

Loan any currently unused property to another government agency. Perhaps a local parks and recreation agency can turn your vacant lot into temporary sports fields or playgrounds until you need the land.

179. Create Community Structures

Find ways to improve properties within your utility boundaries. Part of sustainability is improving the lives of those we serve. If you are able to upgrade any of your sites or areas of your district for the benefit of others, everyone benefits.

180. Promote Employee Volunteerism

Allow employees to volunteer their expertise in water-related issues (improving conservation, fixing leaks, clearing out around meter boxes, and so forth).

Developing a safe, capable, committed workforce and serving customers through community investment are two of Sydney Water's corporate sustainability goals. The Giving through Safety Program brings these goals together. By focusing on the safety of our staff, we generate much-needed funds for key community causes.

Through the Giving through Safety Program, Sydney Water donated over $150,000 to 23 staff-nominated charities in 2007-2008 and $247,000 for 27 charities in 2008-2009. The Giving through Safety Program is now entering its third full year.

The program aims to encourage a stronger safety culture among staff and contractors by setting dollar amounts for specific safety targets. If the targets are met, the dollar amounts are donated to a charity chosen by each team. The program operates in conjunction with other safety policies and procedures to achieve Sydney Water's overall goal of zero injuries.

Contributed by Judi Hansen, Sydney Water

Portland Water Bureau has transformed some of their facilities to create public open spaces called hydroparks. Surrounded by fences, most of these sites were formerly off-limits to local residents. Many were also located in park-deficient neighborhoods. The bureau worked closely with the community to create each of the seven hydroparks. The fences have come down, and in their place the bureau installed benches, walking paths, picnic tables, and dog stations. Two of the hydroparks have community gardens and one has a water conservation demonstration garden. Another of the hydroparks has a community committee working to bring temporary art installations to the site.

Contributed by Kim Dinan, Portland Water Bureau

Northwest Oregon's Regional Water Providers Consortium is a group of 22 water providers in the Portland metropolitan area that directly provide water service to approximately 90% of the urban area. These water providers represent 16 cities, 7 water districts, and 1 people's utility district. The regional government, Metro, not a direct water provider, is also a member, making up a total membership of 25 dues-paying governmental entities.

This organization was formed in 1996 through an intergovernmental agreement and is operated by a set of bodies that meet periodically. The consortium promotes cost-efficient use, wise stewardship, and protection of water resources with the goal of meeting the values of its members and the needs of future generations.

The intent of this organization is to collaboratively discuss, study, and adopt policy and facilitate partnerships to provide municipal drinking water supplies. Though this is a collaborative organization, one of the basic tenets of the consortium is that no individual member gives up its ability to act as it sees fit to provide water services to its customers (RWPC, 2009).

The Portland Water Bureau's Water Event Station (WES) is a portable water station owned and operated by the bureau. WES can be hooked directly into mains and water fountains throughout Portland. WES is offered free of charge at events throughout the city, providing tap water to event participants and eliminating the need for bottled water.

Contributed by Kim Dinan, Portland Water Bureau

181. Collaborate With Other Agencies

Share fueling stations to minimize infrastructure needs. Enlist a parks and recreation agency to chip up your debris for use on their trails. Provide meter reading services for another smaller district. That agency or another may be able to provide bill processing services for you. Look at your region holistically—every agency doesn't have to do everything on a small scale, when economies of scale could be realized by spreading out the tasks to those who can do them well for several organizations.

182. Promote Tap Water in Creative Ways

AWWA's campaigns to promote tap water instead of bottled water can be etched on everything from pint glasses to reusable water pitchers to stainless-steel water bottles. Anything that enables customers to drink tap water more easily is important. Sustainable items such as reusable shopping bags or visors or T-shirts can bear the message, as well.

183. Plant Fruit Trees or Gardens at Utility Properties

Plant trees or gardens at reservoirs, pump stations, or other utility properties. Allow staff time to tend and harvest for the local food bank, or find a local gleaners organization to come pick fruit when ready and give to low-income citizens.

184. Coordinate Construction Projects With Other Entities

To minimize community disruption, proactively communicate with public and private entities to coordinate construction projects. If the county is replacing sidewalks in an area where you are likely to have to replace old pipe with larger pipe or newer material, do the construction at the same time.

This is not a new concept, but it is important to be proactive. Don't just wait until you happen to hear of a project going on in your district. Find representatives from your county, city, electric, natural gas, cable or phone providers, etc. Create a group that meets quarterly or semiannually to discuss what is in the works for each entity. Working with each other to develop schedules will not only minimize community disruption but may also reduce costs for everyone involved.

BAYWORK
Regional Collaborative

Workforce development was one of four topics studied under a Water Research Foundation (WRF) project on the benefits of regional collaboration. The participating California agencies were the Contra Costa Water District, East Bay Municipal Utility District, San Francisco Public Utility Commission, and Santa Clara Valley Water District. The project began in March 2008. As the project was concluding, the utilities (plus the Union Sanitary District) met and decided to form BAYWORK, an informal regional collaborative for continuing workforce development efforts. BAYWORK held its kickoff meeting on June 26, 2009, with invitees from water and wastewater agencies in six Bay Area counties.

Through regional collaboration among Bay Area water and wastewater utilities, BAYWORK seeks to support operational reliability by implementing cost-effective strategies and programs that ensure sufficient numbers of qualified, prepared staff to meet their responsibilities to customers, communities, and the environment.

Objectives

- Get enough of the right people in mission-critical classifications.
- Give staff the information they need to do quality work.
- Modify work processes to optimize use of available staffing.
- Maximize cost-effectiveness of workforce development investments through collaboration.

Activities and Accomplishments

Water/wastewater needs assessment: BAYWORK partnered with the San Francisco Bay Region Center of Excellence, City College of San Francisco, to conduct a needs assessment of mission-critical categories in the water/wastewater business. This definitive study was completed in December 2009 and will help the community college system plan their curriculum and centers of training to fill water industry needs in the future.

Working with workforce investment boards (WIBs): BAYWORK found it very important to be linked with the WIBs as they have the most accurate knowledge of the changing workforce, especially among displaced workers. They worked closely with the Contra Costa County WIB, Alameda County WIB, and NOVA and Work2Future in Santa Clara County.

Working with community colleges: They have worked closely with community colleges, such as Laney College, San Mateo College, Foothill College, Solano College, and the San Jose/Evergreen College. One accomplishment is the development of an electrical and electronic certificate program with the College of San Mateo.

Working in the profession: The work of BAYWORK is coordinated with activities in AWWA, California-Nevada section of AWWA, Water Environment Federation (WEF), and related organizations. Information is now posted on H2Opportunity, a national clearinghouse for career opportunities in the water industry. BAYWORK has also been collaborating with other utilities, such as PG&E, that have similar workforce needs.

(continued)

BAYWORK Regional Collaborative (continued)

Developing the candidate pool: A subcommittee of BAYWORK is working to get more candidates ready to compete for water/wastewater jobs. This includes helping candidates improve math, technical, and communication skills and developing internships and apprenticeship opportunities. The subcommittee is also working on common messaging to publicize careers in the water/wastewater industry to high school and college-level students.

Providing staff with needed work information: A subcommittee of BAYWORK is working to implement videotaping for knowledge retention, sharing best practices on standard operating procedures (SOPs), and video conferencing to share training, talent trading, and mentoring.

Modifying work processes to optimize use of staff: A subcommittee of BAYWORK is working on the use of information technologies (such as using hand-helds), job redesign, and modification of work to match skill sets of employees. See BAYWORK's Roadmap for Regional Workforce Development in appendix N and their needs assessment report at www.coeccc.net/water for more detailed information on their efforts.

Contributed by Ray Yep and Cheryl Davis, BAYWORK

Employee Attraction and Retention

Responsible use of ratepayer and citizen dollars involves hiring and maintaining a highly skilled workforce. The time and money you put into acquiring and keeping your staff involves the sustainability of an asset that is just as important as pipe and trucks. Encouraging employee involvement in a variety of areas and providing training and other benefits is likely a sound, sustainable investment of your customers' money.

185. Provide Continuing-education Reimbursement to Employees

186. Promote Water Utility Careers

It is no secret that the workforce is aging, particularly in the public sector, and many of our employees are retiring with fewer new employees on the horizon. One way to find new hires is to introduce interns from engineering schools to the world of water utilities. Hiring interns will train a few students a year and hopefully gain new applicants for your utility or another. Some high schools have work-study programs that could introduce students to water utility work, as well.

To reach an even larger group, partner with tech schools or community colleges to create certification programs for water-related careers. Consider paying for books and tuition in return for students working for the utility for a certain number of years. Consider volunteer teaching at the school to promote a certification program. The costs of training in this way may be less than training new hires on the job.

Take a regional approach. You can only steal another utility's employees for so long—yours may start going elsewhere, too!

187. Include Your Sustainability Focus in Job Announcements

When you post job announcements, include something like "We are a Sustainably-oriented, SHARP safety graduate, Equal Opportunity Employer" to indicate what is important to your organization. More and more job applicants look for employers that are interested in looking at their work holistically. Promoting your organization's concern for the triple bottom line may encourage just the type of employees you are looking for.

188. Encourage Employees to Provide New Ideas About Sustainability

Offer employee suggestion rewards to foster creativity and obtain new ideas for increased sustainability. If you already have an employee suggestion box, you may be able to just add a sustainability component. Or create a new program specifically for ideas that will make your organization more sustainable environmentally, economically, and/or socially.

You can make it fun, too. A recycled art extravaganza could showcase art made from discarded items. Establish categories for material from work, material from home, useful items, decorative items, or whatever sounds interesting. Get donations from companies that are like-minded or provide sustainable products or services to provide the winners.

189. Consider Alternative Work Schedules

Employ alternative "9/80" or "4 - 10s" work schedules or flextime, giving employees a day off every week or every other week or increased flexibility in travel times. This can keep vehicles off the road at peak times and give employees the opportunity to better fit family needs into their work lives. This has been known to result in fewer absences and increased productivity. It may also reduce carbon emissions if those vehicles are not used on employees' days off, though that can't be guaranteed.

190. Evaluate Employee Ergonomics/Safety

Have an outside professional evaluate employee workstations as needed to provide a safe, healthy working environment. Provide accessories such as foot rests, tilted keyboards, phone headsets, and desk elevators to employees where indicated. Place office equipment at the proper height with appropriate safety features to meet or exceed OSHA standards and protect employees from unnecessary risk.

191. Offer Significant Benefits to Employees

Offer vacation, sick, and comp time; health/dental/vision/life/disability insurance; 401K and other retirement programs; and Section 125 options. Introduce an employee recycling benefit (see above). Once you have spent time hiring and training employees, let them know they are valued and give them ample reason to stay.

192. Provide Services to Employees

Is there anything you can do to reduce the need for trips away from work (for example, dry cleaning pickup and delivery, lunch delivery, or reduced-cost movie tickets)? Many vendors are happy to be able to reach many people at one site. If you can help employees with errands and it doesn't cost your agency more, everyone wins!

193. Keep Employees Well-trained

Training and education of staff and the community are the keys to a water utility's mission. We deliver quality customer service by looking at proactive ideas to increase efficiency, save money, and promote social responsibility among our employees, customers, vendors, and neighbors.

Encourage employees to not only stay up-to-date in their fields but also to expand their skills through cross-training and other means of education. Various options include:

- in-person conferences and seminars,
- teleconferences and webinars,
- Web site self-directed trainings (maybe a series of safety trainings with quizzes to ensure content learning),
- in-house Intranet trainings/PowerPoint,
- library of educational resource material, and
- all-employee meeting presentations including outside speakers.

> *Our aging workforce brings an additional element of risk to the water industry. Each of us has the potential of a major "human asset" going out of service at any time. With the average age of the workforce edging near 50 years of age, the risk to losing the institutional knowledge to manage the infrastructure is huge. Just as we work to make the best investments in our water system and to avoid any disasters within our communities, we need to work to make the best investments in our people and to avoid the disasters that could be created by failing to proactively manage our "human assets."*
>
> (Peterson and Pozos, 2009)

194. Provide Clothes Allowance for Those Staff with Mandatory Uniforms That Allows for Personal Preferences

Personal choice allows employees to look professional while feeling comfortable in their work clothes. Provide organic fabric alternatives, if available. If you use a uniform maintenance company to clean your uniforms, ensure that they are using environmentally friendly products.

195. Engage Employees

Establish a Sustainability Steering Committee, and Wellness, Safety, Green Team, and Employee committees comprised of employees from throughout the organization to encourage participation in decision-making processes. When employees participate in the creation of programs and become more involved with working policies and procedures, they are often happier and more likely to stay with the organization.

The chart on pp. 114–124 recaps the practices described in this chapter. On the right side you will see a breakdown of departments and which groups will likely lead the way in putting these policies and practices into action, although many more employees will participate in and be affected by these efforts.

Sustainability Opportunities at a Water Utility

		MGMT	OFFICE AND SUPPORT SERVICES											ENGINEERING / FIELD OPS						
	POTENTIAL LEAD AREA/ DEPARTMENT (will vary by utility) Many other departments will have key roles and involvement, as well	Executive Management	Administrative Services	Human Resources	Sustainability Program	Conservation Program	Customer Service	Billing	Finance	Purchasing	Information Technology	Partnerships & Public Relations	Building/Grounds Maintenance	Safety	Water Treatment	Distribution	Engineering	Construction	Infrastructure Maintenance	Fleet	Water Quality
	ACTION or PRACTICE																				
	POLICIES AND UTILITY-WIDE INITIATIVES																				
1	Ensure management support of sustainability	✓			✓																
2	Hire a sustainability coordinator	✓		✓																	
3	Adopt sustainability, vision, and/or principles	✓			✓																
4	Develop a sustainable purchasing policy	✓			✓					✓											
5	Include introductory sustainability language and utilize sustainability criteria in RFPs	✓			✓					✓											
6	Integrate sustainability into regular business practices	✓							✓	✓							✓				
7	Add sustainability criteria to capital improvement plan (CIP) evaluations	✓															✓				
8	Establish a green IT initiative	✓			✓						✓										
9	Develop performance measures that address sustainability issues	✓			✓																
10	Develop a carbon emission reduction policy or climate change response plan	✓			✓																
11	Develop a bidder integrity policy	✓								✓											
12	Establish sustainability teams or task forces	✓			✓																
13	Establish a carbon-reduction task force or energy team	✓			✓																
14	Go through a backcasting or similar process to develop targets for sustainable goals	✓			✓																
15	Undertake a process mapping exercise to identify resources used and waste created in agency processes	✓			✓																
16	Conduct an agency-wide carbon emissions inventory	✓			✓								✓							✓	

		MGMT	OFFICE AND SUPPORT SERVICES												ENGINEERING / FIELD OPS						
	POTENTIAL LEAD AREA/ DEPARTMENT (will vary by utility) Many other departments will have key roles and involvement, as well	Executive Management	Administrative Services	Human Resources	Sustainability Program	Conservation Program	Customer Service	Billing	Finance	Purchasing	Information Technology	Partnerships & Public Relations	Building/Grounds Maintenance	Safety	Water Treatment	Distribution	Engineering	Construction	Infrastructure Maintenance	Fleet	Water Quality
	ACTION or PRACTICE																				
17	Adjust all or a portion of your carbon emissions with renewable energy credits and carbon offsets	✓			✓																
18	Join a sustainable networking group	✓			✓																
19	Develop a sustainable food/catering policy	✓	✓																		
20	Incorporate sustainability into human resources policies			✓																	
21	Create an employee recycling benefit	✓			✓																
	ENERGY ISSUES/CARBON EMISSION REDUCTION																				
22	Use natural light and occupancy sensors where appropriate												✓								
23	Participate in an electricity provider program that shows your electricity usage in small intervals												✓								
24	Install a programmable HVAC system												✓								
25	Turn off lights, computers, and electronic devices when not in use				✓						✓		✓								
26	Reduce desktop computer power consumption										✓										
27	Use renewable energy	✓			✓				✓				✓								
28	Use gravity instead of pumping to move water															✓	✓				
29	Install variable frequency drives where appropriate															✓	✓				
30	Optimize pumping systems															✓	✓				
31	Purchase low-emission, two-stroke equipment												✓								
32	Implement a "No Mow" program at appropriate sites												✓								
33	Use goats or other animals to keep remote sites trimmed												✓								
	WATER EFFICIENCY WITHIN UTILITY FACILITIES																				

		MGMT	OFFICE AND SUPPORT SERVICES												ENGINEERING / FIELD OPS						
	POTENTIAL LEAD AREA/ DEPARTMENT (will vary by utility) Many other departments will have key roles and involvement, as well	Executive Management	Administrative Services	Human Resources	Sustainability Program	Conservation Program	Customer Service	Billing	Finance	Purchasing	Information Technology	Partnerships & Public Relations	Building/Grounds Maintenance	Safety	Water Treatment	Distribution	Engineering	Construction	Infrastructure Maintenance	Fleet	Water Quality
	ACTION or PRACTICE																				
34	Install half-flush or dual-flush toilets, waterless urinals					✓							✓								
35	Use low-flow faucet aerators and showerheads					✓							✓								
36	Develop water-efficient landscaping					✓							✓								
37	Use weather-based irrigation systems					✓							✓								
38	Install a rainwater collection and storage system												✓								
39	Employ stormwater solutions that keep water out of storm drains												✓								
40	Drain tanks through distribution system															✓					
41	Appropriately dispose of water from continuous water quality analyzers															✓	✓				✓
42	Eliminate oversized tanks and other inappropriate infrastructure															✓	✓				
	SUSTAINABLE CONSTRUCTION IN NEW UTILITY FACILITIES																				
43	Achieve at least Silver LEED Certification or equivalent	✓																✓			
44	Reduce heat islands to minimize impact on human and wildlife habitat																	✓			
45	Eliminate light escaping from building site to reduce impact on nocturnal environment																	✓			
46	Achieve increasing levels of energy performance above basic building standards																	✓			
47	Include commissioning to ensure building was designed, constructed, and calibrated to operate as intended																	✓			
48	Recycle or reuse maximum construction waste possible–90% or more by weight																	✓			

		MGMT	OFFICE AND SUPPORT SERVICES												ENGINEERING / FIELD OPS						
	POTENTIAL LEAD AREA/ DEPARTMENT (will vary by utility) Many other departments will have key roles and involvement, as well	Executive Management	Administrative Services	Human Resources	Sustainability Program	Conservation Program	Customer Service	Billing	Finance	Purchasing	Information Technology	Partnerships & Public Relations	Building/Grounds Maintenance	Safety	Water Treatment	Distribution	Engineering	Construction	Infrastructure Maintenance	Fleet	Water Quality
	ACTION or PRACTICE																				
49	Use recycled or rapidly renewable products whenever possible																	✓			
50	Use materials manufactured locally whenever possible																	✓			
51	Use low-emitting materials such as carpet, paint, adhesives, and composite wood																	✓			
52	Ensure high indoor air quality through construction process and monitoring																	✓			
53	Install room motion sensors or post prompts by light switches and computers encouraging them to be shut off when not in use												✓					✓			
54	Use integrated pest management												✓								
55	Consider LEED certification for water treatment facilities														✓			✓			
56	Use your building as an example				✓							✓	✓								
	SUSTAINABLE PURCHASING																				
57	Use life-cycle costing (LCC)				✓				✓								✓			✓	
58	Use 100% postconsumer waste (PCW) recycled paper		✓							✓											
59	Purchase office products made from recycled or otherwise more sustainable material		✓							✓											
60	Request office supplies be delivered in reusable bins		✓							✓											
61	Minimize deliveries		✓							✓											
62	Purchase duplexing copiers		✓							✓											
63	Minimize purchase of desktop printers									✓										✓	
64	Minimize purchase of personal heaters									✓			✓								
65	Purchase recycled janitorial products									✓			✓								
66	Use rechargeable batteries									✓											

		MGMT	OFFICE AND SUPPORT SERVICES												ENGINEERING / FIELD OPS						
	POTENTIAL LEAD AREA/ DEPARTMENT (will vary by utility) Many other departments will have key roles and involvement, as well	Executive Management	Administrative Services	Human Resources	Sustainability Program	Conservation Program	Customer Service	Billing	Finance	Purchasing	Information Technology	Partnerships & Public Relations	Building/Grounds Maintenance	Safety	Water Treatment	Distribution	Engineering	Construction	Infrastructure Maintenance	Fleet	Water Quality
	ACTION or PRACTICE																				
67	Purchase Energy Star, Greenpeace-rated or EPEAT-rated computers and equipment									✓	✓										
68	Purchase WaterSense-rated items									✓											
69	Purchase reusable plates/glasses/silverware instead of disposables									✓											
70	If disposables are needed, purchase biodegradable items		✓							✓											
71	Remove bottled water from facilities	✓	✓							✓											
72	Serve "family-style" meeting meals instead of individual servings		✓							✓											
	WASTE REDUCTION/POLLUTION PREVENTION																				
73	Conduct waste audits				✓								✓								
74	Use oil/water separators												✓							✓	
75	Burn waste oil or collect and return for rerefining												✓							✓	
76	Use laptops, scanners, and PDAs as appropriate										✓										
77	Use natural soil-retention products												✓					✓	✓		
78	Make water diversion equipment from discarded bike shop inner tubes and sand																		✓		
79	Use safer water treatment chemicals														✓	✓			✓		
80	Handle and dispose of residuals carefully														✓	✓			✓		
81	Return all unusable metal (cast iron, brass, steel) to the utility for sale as scrap												✓					✓	✓		
82	Use dilute hypochlorine solution at pump stations instead of gaseous chlorine																		✓		
83	Utilize directional boring for laying pipe instead of trenching																	✓			
84	Cover spoils from construction with straw, hay, or jute matting																	✓	✓		

		MGMT	OFFICE AND SUPPORT SERVICES												ENGINEERING / FIELD OPS						
	POTENTIAL LEAD AREA/ DEPARTMENT (will vary by utility) Many other departments will have key roles and involvement, as well	Executive Management	Administrative Services	Human Resources	Sustainability Program	Conservation Program	Customer Service	Billing	Finance	Purchasing	Information Technology	Partnerships & Public Relations	Building/Grounds Maintenance	Safety	Water Treatment	Distribution	Engineering	Construction	Infrastructure Maintenance	Fleet	Water Quality
	ACTION or PRACTICE																				
85	Minimize vegetation removal required at construction sites																	✓	✓		
86	Use lead-free and toluene-free paints for locating services																✓	✓	✓		
87	Separate and recycle concrete and asphalt pieces cut out of streets for leaks and repairs																	✓	✓		
88	Reuse recycled rock and ground tailings in construction projects whenever appropriate																	✓	✓		
89	Practice unidirectional flushing																		✓		
90	Reuse any functional parts from old meters and hydrants to repair other meters and hydrants																		✓		
91	Use reusable aluminum and wood signs at construction and maintenance sites												✓					✓	✓		
92	Partner with local parks and recreation agency												✓								
93	Use fewer toxic materials				✓									✓							
94	Promote online phone books		✓								✓										
95	Request sustainable practices of presenters		✓		✓																
96	Use printed-one-side paper for draft copies or make into scratch pads		✓																		
97	Distribute large documents electronically		✓																		
98	Encourage employee recycling				✓								✓								
99	Use a rubber stamp for incoming payables instead of attaching paper form								✓	✓											
100	Put employee and customer forms online		✓				✓				✓										
101	Implement a paperless plan review process										✓						✓				
102	Develop an automatic electronic archiving system										✓										
103	Put a line at end of e-mails discouraging printing		✓																		
104	Route magazines and documents to interested parties instead of ordering or printing copies for all readers		✓																		

		MGMT	OFFICE AND SUPPORT SERVICES												ENGINEERING / FIELD OPS						
	POTENTIAL LEAD AREA/ DEPARTMENT (will vary by utility) Many other departments will have key roles and involvement, as well	Executive Management	Administrative Services	Human Resources	Sustainability Program	Conservation Program	Customer Service	Billing	Finance	Purchasing	Information Technology	Partnerships & Public Relations	Building/Grounds Maintenance	Safety	Water Treatment	Distribution	Engineering	Construction	Infrastructure Maintenance	Fleet	Water Quality
	ACTION or PRACTICE																				
105	Foster innovation				✓																
106	Compost coffee grounds, fruit and vegetable scraps, tea bags, etc.												✓								
107	Use nontoxic cleaning supplies												✓								
	REDUCE/REUSE/RECYCLE																				
108	Increase recycling awareness with education and signage				✓								✓								
109	Create an employee recycling center				✓								✓								
110	Send everything possible out for reuse or recycling including items 111–136				✓								✓								
111	alkaline batteries (from pagers, locaters, cameras, cell phones, etc.)												✓								
112	block Styrofoam												✓								
113	CDs/DVDs												✓								
114	cell phones												✓								
115	construction debris																	✓			
116	electronic items										✓		✓								
117	floppy disks												✓								
118	glass												✓								
119	lawnmowers												✓								
120	light and corrugated cardboard												✓								
121	light bulbs												✓								
122	metal (copper, aluminum, iron, tin)												✓								
123	meter registers												✓						✓		
124	motor oil												✓								
125	newspaper												✓								
126	ni-cad batteries												✓								

		MGMT	OFFICE AND SUPPORT SERVICES											ENGINEERING / FIELD OPS						
	POTENTIAL LEAD AREA/ DEPARTMENT (will vary by utility) Many other departments will have key roles and involvement, as well	Executive Management	Administrative Services	Human Resources	Sustainability Program	Conservation Program	Customer Service	Billing	Finance	Purchasing	Information Technology	Partnerships & Public Relations	Building/Grounds Maintenance	Safety	Water Treatment	Distribution	Engineering	Construction	Infrastructure Maintenance	Fleet	Water Quality
	ACTION or PRACTICE																				
127	packing air bags and packing peanuts												✓								
128	paper												✓								
129	plastic bags												✓								
130	plastic bottles and other plastics accepted by your hauler												✓								
131	printer/toner cartridges												✓								
132	telephone books												✓								
133	tennis shoes												✓								
134	uniforms												✓								
135	useable shoes												✓								
136	wood pallets												✓								
	TRANSPORTATION ISSUES (employee commute trips, customer trips, company fleet operations)																				
137	Implement LEED features such as carpool spaces, bike storage, employee showers												✓								
138	Trip planning		✓		✓																
139	Establish satellite offices or check-in sites	✓																			
140	Require reduced vehicle idling in shop yard and at work sites																			✓	
141	Make proper vehicle maintenance a priority																			✓	
142	Evaluate your vehicle purchases									✓										✓	
143	Use alternative fuels—biodiesel, ethanol, etc.—as appropriate									✓										✓	
144	Examine your oil usage practices									✓										✓	
145	Use the appropriate vehicle for the job																			✓	
146	Coordinate errands		✓																		
147	Provide free or discounted transit passes for all staff members	✓		✓																	

		MGMT	OFFICE AND SUPPORT SERVICES												ENGINEERING / FIELD OPS						
	POTENTIAL LEAD AREA/ DEPARTMENT (will vary by utility) Many other departments will have key roles and involvement, as well	Executive Management	Administrative Services	Human Resources	Sustainability Program	Conservation Program	Customer Service	Billing	Finance	Purchasing	Information Technology	Partnerships & Public Relations	Building/Grounds Maintenance	Safety	Water Treatment	Distribution	Engineering	Construction	Infrastructure Maintenance	Fleet	Water Quality
	ACTION or PRACTICE																				
148	Establish a free ride home program	✓		✓																✓	
149	Provide biking amenities												✓								
150	Maintain alternative sites for bill payment						✓	✓													
151	Provide online bill notification, bill payment, and payroll options						✓				✓										
152	Provide an electronic funds transfer (EFT) payment method to reduce travel and paper usage							✓			✓										
153	Utilize teleconferencing										✓										
154	Schedule meetings to minimize driving	✓	✓																		
155	Provide incentives for meeting attendees to carpool or use public transportation				✓							✓									
156	Challenge, encourage & recognize employees who use alternative transportation		✓		✓															✓	
157	Allow flexible work hours	✓		✓																	
158	Share a fueling station with another agency and agree to some percentage of biofuels																			✓	
159	Participate in community events to cut down on single-person commuting		✓		✓							✓									
	OUTREACH AND EDUCATION																				
160	Provide water conservation training to customers					✓	✓														
161	Educate customers regarding water conservation					✓	✓														
162	Reduce resource demand through water conservation programs					✓															
163	Adopt a pricing structure that encourages water conservation	✓							✓												
164	Advocate for water supply obtained from sustainable sources	✓			✓												✓				
165	Encourage adoption of sustainable practices among customers, suppliers, and/or peers				✓							✓									

		MGMT	OFFICE AND SUPPORT SERVICES												ENGINEERING / FIELD OPS						
	POTENTIAL LEAD AREA/ DEPARTMENT (will vary by utility) Many other departments will have key roles and involvement, as well	Executive Management	Administrative Services	Human Resources	Sustainability Program	Conservation Program	Customer Service	Billing	Finance	Purchasing	Information Technology	Partnerships & Public Relations	Building/Grounds Maintenance	Safety	Water Treatment	Distribution	Engineering	Construction	Infrastructure Maintenance	Fleet	Water Quality
	ACTION or PRACTICE																				
166	Make sure incentive items are useful and sustainable				✓	✓															
167	Train new employees in sustainable practices			✓	✓							✓									
168	If you have a speakers bureau, include sustainability experts in your offerings				✓							✓									
169	Make presentations regarding your progress to organizations and trade associations	✓			✓							✓									
	COMMUNITY SERVICE AND CIVIC INVOLVEMENT																				
170	Create an emergency assistance program for customers						✓	✓													
171	Build underground storage tanks when possible to minimize environmental impact and maintain natural look of area																✓				
172	Participate in neighborhood litter cleanup program or other community betterment event		✓									✓									
173	Allow employees company time to volunteer	✓		✓																	
174	Encourage employee-supported donations for local nonprofit groups	✓		✓	✓							✓									
175	Participate in conservation programs, educational fairs, and other community events as appropriate				✓	✓						✓									
176	Take into account minority/women-owned and emerging small business status in purchasing									✓											
177	Produce a green team resource book available for employee use				✓																
178	Share unused property	✓																			
179	Create community structures	✓			✓	✓							✓					✓			
180	Promote employee volunteerism	✓		✓								✓									
181	Collaborate with other agencies											✓									
182	Promote tap water in creative ways					✓						✓									

		MGMT	OFFICE AND SUPPORT SERVICES												ENGINEERING / FIELD OPS						
	POTENTIAL LEAD AREA/ DEPARTMENT (will vary by utility) Many other departments will have key roles and involvement, as well	Executive Management	Administrative Services	Human Resources	Sustainability Program	Conservation Program	Customer Service	Billing	Finance	Purchasing	Information Technology	Partnerships & Public Relations	Building/Grounds Maintenance	Safety	Water Treatment	Distribution	Engineering	Construction	Infrastructure Maintenance	Fleet	Water Quality
	ACTION or PRACTICE																				
183	Plant fruit trees or gardens at utility properties												✓								
184	Coordinate construction projects with other entities																	✓	✓		
	EMPLOYEE ATTRACTION AND RETENTION																				
185	Provide continuing-education reimbursement to employees			✓																	
186	Promote water utility careers			✓								✓									
187	Include your sustainability focus in job announcements			✓																	
188	Encourage employees to provide new ideas about sustainability				✓																
189	Consider alternative work schedules	✓		✓																	
190	Evaluate employee ergonomics/safety													✓							
191	Offer significant benefits to employees	✓		✓																	
192	Provide services to employees			✓																	
193	Keep employees well-trained			✓																	
194	Provide clothes allowance for those staff with mandatory uniforms that allows for personal preferences						✓														
195	Engage employees in sustainability	✓																			

Resources

Desk Heaters

- The University of Sydney, http://www.usyd.edu.au/ohs/policies/ohs/heaters.shtml
- Electronic Product Environmental Assessment Tool (EPEAT) rankings, www.EPEAT.net
- Greenpeace International Guide to Greener Electronics, http://www.greenpeace.org/international/campaigns/toxics/electronics/how-the-companies-line-up

Green IT Resources

- Greener Computing, www.greenercomputing.com
- The Green IT Review, www.thegreenitreview.com

The Green Grid (Data Centers), www.thegreengrid.org.

- Climate Savers Computing, www.climatesaverscomputing.org
- Smart 2020 (A research report with analysis on how information and communications technology could reduce emissions. Sponsored by The Climate Group.), www.smart2020.org
- Velte, T., A. Velte, and R.C. Elsenpeter. 2008. *Green IT: Reduce Your Information System's Environmental Impact While Adding to the Bottom Line.* New York: McGraw-Hill.
- Baroudi, C., J. Hill, A. Reinhold, and J. Senxian. 2009. *Green IT For Dummies.* New York: John Wiley & Sons.
- Lamb, J. 2009. *The Greening of IT: How Companies Can Make a Difference for the Environment.* Indianapolis, Ind.: IBM Press.

Greenhouse Gas and Energy Co-benefits of Water Conservation

- Polis Project on Ecological Governance, http://www.poliswaterproject.org/publication/91

Integrated Pest Management

- Wikipedia: Integrated Pest Management, http://en.wikipedia.org/wik/Integrated_pest_management
- National Sustainable Agriculture Information Service: Biointensive Integrated Pest Management (IPM) Fundamentals of Sustainable Agriculture, http://attra.ncat.org
- Benbrook, C.M. 1996. *Pest Management at the Crossroads.* Yonkers, N.Y.: Consumers Union.

International Society of Sustainability Professionals (ISSP)

- www.sustainabilityprofessionals.org

(continued)

Workforce Planning

- Isbell, M., et al. 2008. Recruit and Retain Great Employees: Proven Approaches You Can Implement Now to Address the Workforce Gap. *Jour. AWWA*, August, 2008. http://apps.awwa.org/WaterLibrary/showabstract.aspx?an=JAW_0070532.
- BAYWORK. 2009. Building and Recruiting Qualified Candidates for Water Industry Jobs. *Jour. AWWA*, August, 2009. http://apps.awwa.org/WaterLibrary/showabstract.aspx?an=JAW_0070538.

REFERENCES

Barwon Water. 2009. http://www.barwonwater.vic.gov.au/cms/serveDoc.cfm?docId=21123. Accessed 1/9/10. Personal communication with Jeremy Rann (Corporate Communications Officer, Barwon Water) on 1/19/10.

Bonneville Environmental Foundation. 2009. http://www.b-e-f.org/faqs/general. Accessed 1/2/10.

City of Boulder. 2009. http://www.bouldercolorado.gov/index.php?option=com_content&task=view&id=7516&Itemid=392. Accessed 1/9/10.

Conrad, S. 2009. Private communication from S.A. Conrad on 1/18/10.

Dinan, K. 2009. Personal communication with Kim Dinan (Sustainability Coordinator, Portland Water Bureau).

Free Geek. 2009. www.freegeek.org/. Accessed 11/13/09.

GOOD Web site. 2009. http://awesome.good.is/transparency/008/trans008vampireenergy.html. Accessed 1/4/10.

Hansen, J. 2009. Personal communication with Judi Hansen (General Manager, Sustainability, Sydney Water) on 12/20/09.

ILBI (International Living Building Institute). 2009. http://ilbi.org/. Accessed 1/2/10.

International Society of Sustainability Professionals (ISSP). 2009. www.sustainability-professionals.org/. Accessed 1/10/10.

MSNBC. 2009. http://www.msnbc.msn.com/id/5279230/. Accessed 1/9/10.

Peterson, J., and N. Pozos. 2009. Sustaining a Workforce With the Skills, Knowledge, and Talent Required to Succeed. *Jour. AWWA*, 101:8:35. http://apps.awwa.org/WaterLibrary/showabstract.aspx?an=JAW_0070532.

Regional Water Providers Consortium (RWPC). 2009. Compiled from Web site and 5-year Strategic Plan. www.conserveh20.org. Accessed 7/11/09.

Seattle Public Utilities. 2009. http://www.seattle.gov/DPD/GreenBuilding/CapitalProjects/DesignToolsStrategies/DPDS_007481.asp#RFQ. Accessed 1/9/10.

Tualatin Valley Water District. 2009. District RFP language.

USEPA. 2009. Integrated Pest Management (IPM) Principles Fact Sheet. Washington, D.C.

Wolan, M., S. Conner, and T. Kuster. 2007. Harnessing the Sun: Solar Panels Power Treatment Plant. *Opflow*, 33:5:10. May 2007.

ZWA (Zero Waste Alliance). 2009. http://www.zerowaste.org. Accessed 11/13/09.

Chapter 5

How Do We Talk About Sustainability?

In addition to getting things done, we also have to figure out how to explain and sometimes justify what we're doing, get support for those actions, and encourage others to join us. The communications required are different for each audience and may be different for the same audience at different times. Virtually all such communications will involve presentation of data, but some will require the art of persuasion, as well. All will entail good writing or speaking skills, but they will also require you to listen. Sustainability is really about organizational change, and change is not easy for most people. How we talk about what we do can often make or break a project or proposal.

The most important thing is to speak with passion. Behaving sustainably is essential to your utility's success. It is easy to feel you are adding to someone's workload or budget by introducing a new way of thinking about how hydrants are flushed or asking the fleet manager to look for a more fuel-efficient truck. But these products and processes are the foundation of your organization. They deserve to be reexamined for increased efficiency and innovation just as you regularly look for the best price on pipe or paper clips. Revisit chapter 2 to recall how beneficial sustainability can be to your organization. By encouraging sustainability you improve all three of the bottom lines your organization is concerned with. Make sure you get those points across.

Know what is important to your audience, whoever that is. I try to use the word *sustainability* as little as possible when talking with people about that very subject. Through one-on-one discussions or from just being observant you might see that manager A is really excited about his grandkids or coworker B is enthusiastic about how efficient she can be when changing out a meter. Use these observations to make points that resonate with each person whenever possible. What you are proposing must be relevant to the audience or you've lost them.

Show them, too, that you understand that sustainability is a part of the whole, not the whole picture. The wonderful thing about sustainability is that it takes into account not only budgets and economic realities but also community desires, political pressure, and environmental, health, and safety concerns. It is a complex balance, and not an easy one to master. But let people know that you are aware that they have a lot to consider as they make decisions in their work and home lives.

Know what you want from the session, be it a discussion with the general manager or a presentation to your entire staff. Prepare your case with points you want to make and messages you want to drive home. Sometimes these meetings are short and it is good to know what you want to include.

And know what you want to ask from your audience. Do you want someone to be a resource for others? To take on a sustainability-related duty? To adopt a sustainability resolution? To pledge to turn off their computers at night? Know the outcome you want and ask for it!

Employee Communications

Employees are your sustainability ambassadors. You can't be a one-person campaign. These are the people who will take the message to others and bring you information about how the organization can make its operations more sustainable. After a while, they are the ones who will start making changes on their own, knowing that a more sustainable utility is more efficient, more effective, and better able to face future challenges in all areas.

Employees generally need material to be "user-friendly," fun, and interesting. Tell them what they can do without telling them what to do and how to do it. Give them information without nagging. It's a fine line. As with most communication, figure you're talking to the 80% who are most interested in listening. Most of the other 20% will likely come along later, once the majority has adopted a culture of sustainability.

Some examples of places to engage employees regarding sustainability are discussed in the following sections.

All-employee Meetings

If you have a small enough staff that you all meet together at regular intervals, this is the perfect place to introduce outside speakers, show PowerPoint presentations or DVDs, explain new policies, stage skits, even have discussions about ways to become a more sustainable utility. If you never get together in one large group, then perhaps you can make similar presentations to several crews or departments at one time.

There is great value in having a common language and shared experiences. Anything you can do to bring information to large parts of your staff at once will be helpful. This is especially important if you have a speaker come to talk about foundational issues, for example, training the staff on The Natural Step or rolling out a new recycling program. Even if you can have half your staff get the training in the morning and the other in the afternoon, it is good to get the momentum and "buzz" going among everyone at the same time.

Two sample formats that have worked well for me in these larger meetings are illustrated below. One is a game show type of design. In this case it deals with recycling. What do you think we can recycle at this utility? Bring each item up separately and after you get the answer from the crowd, show the yes or no button. To spice it up a little, throw candy or something more useful to the first person you hear say the correct answer. Add caveats if needed for a particular item and stop to have a brief discussion if someone has a question (see Figure 5.1).

It's so easy to say yes or no, while remaining fairly anonymous, that most people will participate at one point or another. Start with the easy answers and work up to items that you wouldn't think could be reused or recycled. My favorite is lawnmowers. We have a place in our area that refurbishes them and resells them. They are happy to get the mowers and our employees are happy to have a place other than the dump to take their old ones. But the first time we played this game, most people had no idea we could recycle lawnmowers.

Another format that worked well was the "virtual tour." When we reintroduced our recycling program, the green team wanted to take the staff around and show them all the recycling areas. There are about 25 different places to recycle items at our headquarters. We knew that 110 staff members would not trail around after a leader to all of those sites, so we put together a virtual tour of the facility.

Our green team chairs went through the building taking pictures of what was recycled and where. For example, "Here's the battery bucket by the Water Quality staff area. Around the corner and out the door you'll find the employee recycling center where you can recycle glass, packing peanuts, cell phones, CDs, etc." (See Figures 5.2 and 5.3.) Be very specific and include as many staff members in the photos as you can to keep everyone's interest!

Small Groups

Different opportunities are available when you talk with smaller groups. Meeting with individual crews or departments allows participants to give you more candid feedback than you might get in a larger group. You can ask which

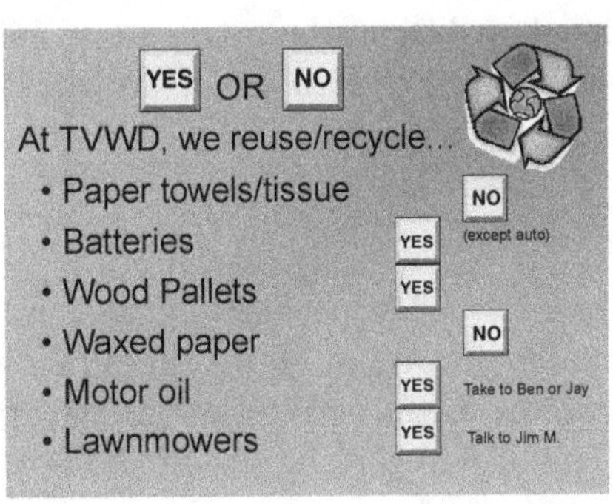

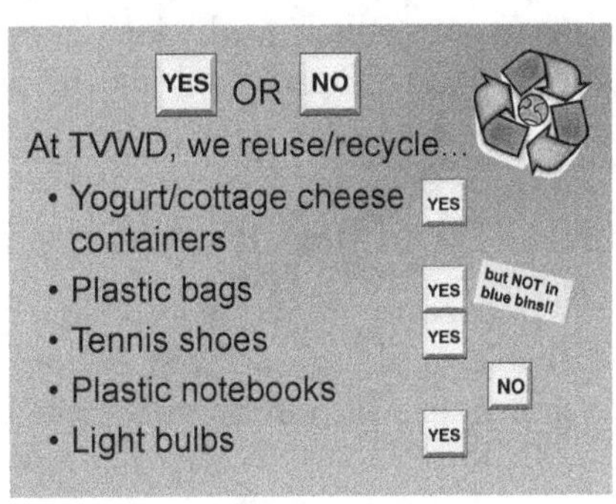

Figure 5.1. Explain new recycling policies and other sustainability issues using staff presentations like the ones shown here

Figure 5.2. Batteries collected here!

sustainability initiatives are going well and which need some work or maybe need to be scrapped. Employees will know more about their own jobs than you could ever learn, and they are the best people to tell you where opportunities exist to improve sustainability in their areas.

If you are able to devote a sizeable amount of staff time to a short-term sustainability effort, consider the process mapping exercise described in chapter 3. This is a great way to glean information about (1) resources used in all the processes in which your utility is involved, (2) waste that might be utilized in other ways, and (3) communication issues and other miscellaneous interactions between groups within your organization. Though temporarily time-consuming, it is a valuable way to look at your utility holistically and discover opportunities for improvement that separate groups would never find.

As you can, arrange one-on-one meetings, too. Listening to what is important to a manager (or any coworker) will inform your interactions with them. One manager's focus may be entirely on staying within budget. You know to start there with cost-saving projects—energy efficiency or waste reduction—and branch out after those projects are successful. If another's priority is reducing hazardous materials to provide a safer working environment, work with this person to create an inventory of hazardous substances used and help him or her find alternatives that do the job. These people will likely become sustainability advocates in the future, and your time and interest in *their* priorities will have been well spent.

Senior Management Meetings

You will need to get buy-in from all of your managers at some point to embed sustainable thinking into your culture as an organization. How you do this will vary depending on your organizational structure. Gaining access to all managers at once allows you to develop a common language regarding sustainability concepts and hear senior management issues and questions so you can get a larger view of how the organization will be affected in any given situation. Seeing "the big picture" is essential to making sustainable solutions work for everyone.

Figure 5.3. Employee recycling center makes it easy to recycle

It is also helpful to look at issues at the department level. Find examples in each manager's department where products or processes that could be considered sustainable are already in place. Look at the table in chapter 4 to see which of the activities listed are already happening. There will be more than you expect! Point out how far you have already come and what natural steps are next.

Budget committee meetings are important places to have a presence. Depending on your structure, you may talk to managers one-on-one about sustainable purchases or there may be budget request forms or smaller committees to address. Until sustainability is more ingrained in the utility's culture, if you don't introduce and advocate for sustainable products and processes, then some managers might not think about it.

Newsletters

Whether printed and stuffed in a paycheck envelope or sent out electronically, newsletters can be a good avenue to reach some employees. There will likely be some staff who read the communication top to bottom and others who will never see it. Don't count on just a newsletter to get out anything important, but it is a good tool when used with others. Use multiple messages in different formats to really make your point.

It can be time-consuming to find information for a weekly or monthly newsletter. A few ideas to ease the burden follow.

- Rotate the duty between green team members or others interested in writing up information or tips for employees.
- Coordinate with other agencies or businesses that also want to get information on sustainable activities out to their employees. You may have to write something just once every few months if you get enough participants to share the load.

Tap into online blogs and electronic newsletters (e-newsletters) whose staff members spend their time finding interesting information and tips relating to environmental and sustainable living. It's usually fine to reprint these in full or edit them down, as long as you give the author proper credit. Search the Internet under "weekly sustainable tips" and you will find plenty of information to share with your coworkers.

Note that not everything you share will be appreciated by your readers. Some readers will think some tips are too simple to be useful; others will find more "advanced" tips to be out of their reach and ridiculous. As with this book, all we can do is put information out there and trust that, though all the ideas will not be appropriate for everyone, each person will find a few helpful tidbits from time to time.

The downside of this is that the articles can't be as tailored to your employees as they can if all of them are written by you or your own employees.

E-mails

Aside from regular communications, carefully chosen e-mails may be helpful, too. Information such as a community recycling event that takes miscellaneous plastics you can't recycle through your utility, notice of an award or an employee's outstanding work on getting solar panels on the roof, or a hybrid in the yard might be worth a quick e-mail.

Posters

Here is a format that brings up a question of sustainability. Is it worth using resources, such as paper, ink, and the electricity to produce posters, to raise awareness about sustainability? One could ask the same thing about using electricity to send e-newsletters, but somehow print media usually gets the scrutiny. There's no right answer, of course. If there are issues that need to be in front of people, especially those who are not in front of computers all day, paper and ink may be the solution.

An example of effective use of posters might be an anti-idling campaign. You can make a presentation to your full staff, you can meet with field crews, and you can talk to individuals as needed. But some well-placed 8.5 × 11 signs or larger posters may keep the issue in employees' consciousness and result in better participation. It is a judgment call, as so much of this work is. But it is better to have thought about the possibility of perceived hypocrisy before that charge is leveled.

Keep the focus on taking the next step—wherever and whatever that is. We do use resources in our work, for example, pipe, computers, and paper, but our goal is to think about our resource usage and use what we need efficiently. Given

> TIP: You know how many e-mails you get each day. Make sure when you send an e-mail it is important or yours will start being ignored.

TIP: If you're not sure what type of recognition programs are available, there may be something like the EcoMetro Guide (www.ecometro.com) in your area that lists businesses that are making efforts to be more sustainable. That could give you some ideas. Or SustainLane's discount coupon section (http://www.sustainlane.com/green-saver-discount-coupons/) may help.

FOLLOW-UP TIP: So as not to cause employees any problems, check out any applicable restrictions on how much public employees can legally receive in gifts from a particular vendor and whether or not these gifts are taxable.

that, signs or posters used judiciously can be an appropriate use of resources in some circumstances.

Recognition

If your organization has an employee suggestion program, start recognizing employees for sustainable submissions, too. If you do not have such a program, consider implementing one, specifically for sustainability or for a broader range of suggestions. This might be in the form of a "pat on the back" for a suggestion that improves energy efficiency or reduces carbon emissions. It might be a token financial reward for a suggestion accepted by a review team or maybe only for one that will save the utility money. You will have to decide on the parameters that work for your organization.

Or you might start a "Do the Right Thing" campaign with small rewards given out for employees "caught" improving the sustainability of your utility. Members of your green team, supervisors, or managers could be given the authority to write up any staff member who has gone above and beyond to, for example, set up a new recycling area, work with other staff to make sure their vehicles aren't idling too long, or research new products that are less toxic… the list is endless. To keep the recognition fair, make sure that workers in the field have as much opportunity as office workers to be seen doing good things.

You could ask local sustainable businesses to donate small prizes as rewards. These might include a gift certificate to an organic food store or restaurant, a tune-up at a bike shop, or a coffee card at a shop that sells fair-trade, shade-grown coffee. Look around and see what is in your community.

Vendor/Organization Fairs

Invite people from your community to offer resources for employees. These could include representatives from your waste-hauling company, recycling facility, electric utility, local food co-op, local sustainable businesses, farmers, irrigation district, sustainable foresters, LEED-certified or Living Building Challenge architects, and solar installers. Think about who has good information to help employees live more sustainably.

Nonprofits dealing with waste reduction, sustainable food options, building, purchasing, or sustainability are, in general, always anxious to share their work with others. Do an Internet search for "sustainability nonprofits," "sustainability NGOs," or "sustainable businesses" followed by your city and state or province. Or simply search for "sustainable (your city here)." Governments also will likely have resource people for recycling or energy conservation, for instance.

Invite representatives to come show their products and maybe provide a short description of their wares or services. Do this over a lunch hour, during an all-employee meeting, or in conjunction with another already established trade show or fair showcasing things like safety items or new equipment.

Brown Bag Lunches/Discussion Groups

Another venue for allowing presentations from some of the above groups is a brown bag lunch. Employees can bring their lunch and listen to a speaker on their own time. Employees might also be interested in a book group in this format, where they can discuss sustainability-related books. Some organizations have already structured discussion groups that are appropriate for lunch or after-work time.

The Northwest Earth Institute (NWEI), for example, provides a series of discussion courses available for small groups anywhere in North America. The courses provide an enjoyable, supportive setting in which to examine personal values and habits, engage in stimulating conversation, create meaningful community, and consider ways to take action toward creating a more sustainable future.

Their courses have such titles as: "Sustainable Systems at Work," "Global Warming—Changing CO_2 Use," "Choices for Sustainable Living," "Menu for the Future," "Voluntary Simplicity," and "Healthy Children/Healthy Planet." NWEI discussion course study guides are manuals for a self-facilitated course experience, taken informally by small groups in a home, at your workplace, a center of faith, or any place where people naturally gather (NWEI, 2009).

Check universities, community colleges, and other centers in your area for study groups, too.

Contests and Challenges

Participation is key to behavior change, and contests can be a fun way to involve staff in thinking about things in new ways. Promote a recycled art contest to encourage employees to think about waste in a different way. You might establish categories for material from work (meter registers, scrap pipe, wood pallets, etc.) and material from home (corks, plastic bottles, metal caps, etc.). You could offer prizes for useful items, decorative items, or whatever sounds interesting. Get donations from companies that are like-minded or provide sustainable products or services to the winners. Involving board members or management in the judging might be a good way to introduce those folks to the concept of waste reduction. Or allow all staff to vote on the winners.

Contests to create logos or posters for various events might be appealing to the engineers and artists in your midst. And everyone can participate in suggesting a new name for the green team or naming an event. Anything you can do to get people thinking about sustainability and acting in some small way will help ingrain new ways of thinking.

A sustainable living challenge could be a good way to illustrate the negative impact people have on the environment and what they can do to reduce it. Start by giving everyone a Web site where they can assess the impact they make by the way they currently live.

One option is the impact calculator at www.lowimpactliving.com. I like this one because it assesses not only carbon emissions, as many do, but also water and energy usage, wastewater, runoff, and waste generation. It is not perfect and you will likely get questions from some participants about how to calculate the size of their lot or how to record the fact that they only offset part of their car's emissions. It won't precisely fit everyone's situation. The point is to think about our lives and how we might modify our actions to reduce our negative impact on the environment and increase our positive impact within our community. The precise score is not the whole picture.

Another popular calculator is www.myfootprint.org. It takes a more global view but doesn't offer quite as many suggestions for how to reduce your footprint. It is, however, available in five different languages and offers lots of education throughout the quiz. And it is humbling to find out that even if you are using half as many resources as your country as a whole, you may still be using three or four times more than the earth's carrying capacity. That is, if everyone on earth lived your lifestyle, then we would need three or four earths! Good information to have, especially if you think you're already doing a great job of living sustainably!

Other calculators I've run across are listed in the Resources section at the end of this chapter.

Next step. When each person has taken whatever quiz you decide on, have them send their score to one person for compilation. Post the average for your utility or work group. That establishes a baseline.

The important part is to then conduct an educational campaign, bringing in speakers or developing presentations to show employees what they can do to use fewer resources and create less waste.

- Your electrical utility representative may be willing to come in to talk about energy efficiency or even solar, wind, or other renewable energy opportunities for residential customers.
- Someone from your own water conservation program could talk about efficient water usage.
- Members of a farmers' market or food co-op could discuss local organic food options.
- An auto dealership that specializes in hybrids or alternatively fueled vehicles may have someone willing to tell your staff about the latest in more sustainable vehicle technologies for cars and trucks.

> TIP: If you have multiple locations, competition between various work groups might be fun, but remember that this is a self-assessment, so manipulation is possible and the results are not scientific. It is, again, one way to draw people in, get their active participation, and help them learn something they can use.

The possibilities are endless, and your community is no doubt filled with businesses, nonprofits, and individuals willing to help educate your employees. This program could take six months to a year.

As part of this educational push, a vendor or organizational fair, as described above, may also give employees access to helpful new resources.

The final step. Have participants take the same online assessment after the educational program is complete. See what the average score is now from the participants who took the initial assessment and discover whether or not the additional education made a difference. It might not have, but just the increased awareness is likely to have encouraged staff to adopt new behaviors and purchase (or not purchase!) different products.

Process Mapping

Process mapping was discussed in detail in chapter 3. It can take up a fair amount of staff time but is a great way to get input from employees and assess the utility as a whole.

Elected Official Communications

If your elected officials have work sessions that are open to the public, but are more informal than their regular meetings, then these can be great ways to educate these folks about various topics.

This is often your best opportunity to have a dialog with your elected officials. Some issues, such as power purchase agreements or renewable energy credits, require some explanation and deserve some time spent answering questions. If you can get this more informal time with them, by all means take advantage of it. Time with individual officials is valuable, too, but it is even better to develop the common language with them as you did with management.

Many of the presentations you might have given for employees in formats described above can be modified to focus on what elected officials are concerned with. While employees are basically concerned with their own behavior, elected officials will be more concerned with how utility policies and staff actions affect citizens and ratepayers. Refer to chapter 2 to find the parts of the business case that will make sense to each of them.

Your other opportunity to speak to elected officials will likely be in formal board meetings, where proposals will be approved and resolutions adopted. By this time your officials need to understand the issues and hopefully be counted on to support your suggestions. Any presentation will likely be for the benefit of the public, whether in the room, listening to streamed audio, or watching a televised meeting. This requires a concise description of the issue and the action desired by the elected officials.

If there is a regular written (probably e-mail) communication that goes to commissioners, you may have opportunities to write white papers or share information of value to your officials. This is a good way to impart information that requires graphs or tables or may need to be reread or reviewed closely.

Some managers are hesitant to talk about an organization's performance if it is not excellent. We all want to tell our elected officials, customers, and the general public about what is working well. The San Francisco Public Utility Commission (SFPUC) has on their Web site an excellent PowerPoint presentation that tells the whole story while focusing on the work that is being done to improve performance in various areas.

They present their sustainability plan and program as: "A system for evaluating our department-wide business performance that takes into account the financial, environmental, and social impacts of our business activities." The presentation goes on to provide definitions of sustainability, who is involved, and how they go about evaluating performance. They then show several "Sustainability in Action" slides (Figure 5.4). This puts the emphasis on the process, not the exact scores received in any area. SFPUC is a leader in water utility sustainability efforts, but in a few areas they just weren't where they wanted to be. In one case, it was simply because they had not been measuring something they thought they should be. They portray the organization in a positive light but let viewers know SFPUC is still striving to improve and detail what action is being taken to do that.

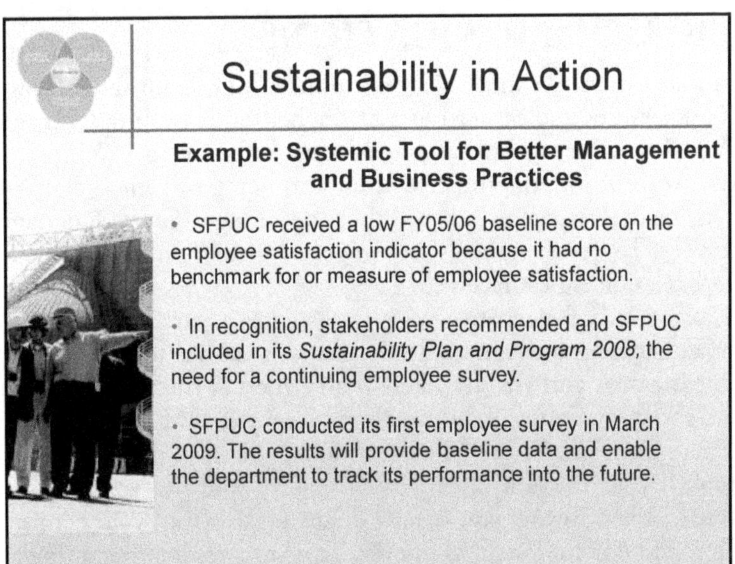

Figure 5.4. Slide describing sustainability accomplishments and opportunities

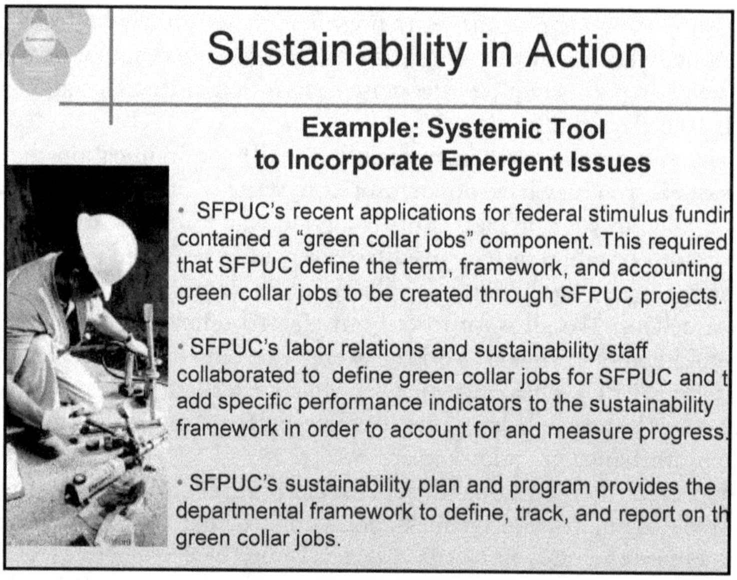

Figure 5.5. SFPUC highlights its forward-looking approach

Other highlighted actions, such as Figure 5.5 describing a systemic tool to incorporate emergent issues, show innovation and showcase SFPUC's forward-looking approach. Most elected officials like to know that their organizations are thinking ahead and will not only be prepared for upcoming regulation or legislation but are also interested in doing well by all their stakeholders.

PUBLIC COMMUNICATIONS

There are many ways to contact your customers. Figure 5.6 shows how Miami-Dade County Water and Sewer District reaches out regarding their rebate programs. You'll see that their Web site is most effective in this situation, with brochures, word of mouth, and television also being of some help. Note, also, that the radio advertisements reached those speaking different languages to different degrees. To get the greatest coverage, pay attention to cultural variations in how information is best received.

This example is for rebates, which are likely a part of your conservation program. Publicizing our conservation efforts is crucial to utilizing water in a sustainable manner, and you are likely pretty good at that already. Telling your customers about your sustainability efforts is a little different. Sustainability is not your main mission. You don't exist simply to be sustainable, but you do exist to accomplish your mission of providing quality water to your customers in a sustainable manner. Even if you haven't defined it that way, your commitment to the public is to be there for them forever. That involves looking at how you provide water and how your processes may need to be modified to do that in a way that is most cost-effective, least environmentally harmful, and most beneficial to your community.

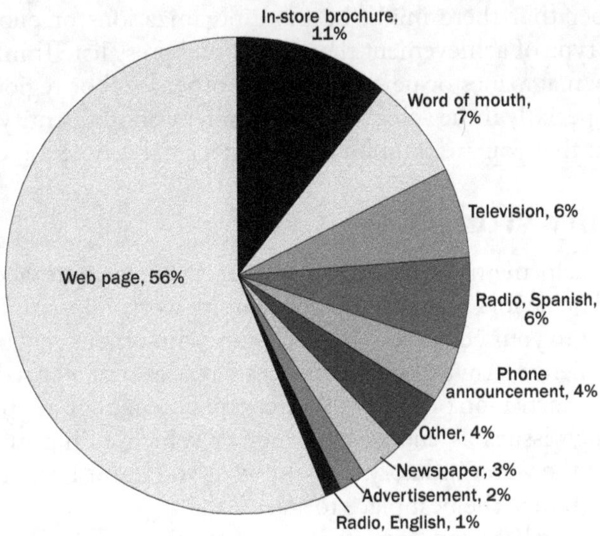

Figure 5.6. Public outreach for high-efficiency rebate programs

Press Releases

Including sustainability in how you talk about your central mission is often the way it is best received. When you tout just your efforts to be more sustainable, the public can read that as spending more money than you need to. When you show how it makes your operations more reliable, less costly, or safer, it's usually seen as a win for everyone.

- Are you installing solar panels through a power purchase agreement that will cost you nothing but will take the uncertainty out of your power costs and allow you to purchase renewable instead of petroleum-based power for a particular facility?
- Have you installed variable frequency drives at pump stations that have reduced your electricity costs?
- Have you put a micro hydro-facility online that creates renewable energy from your water and pays you for that generation?
- Did you complete an inventory of hazardous substances in your facilities that resulted in consolidation of products (less waste), disposal of unused products (less risk of spills and leaks), or substitution of new, less toxic substances (increased employee health and safety)?
- Have you gotten an award or certification or was a case study done on your utility for community service?
- Have you completed a LEED- or Living Building Challenge-certified facility that used less toxic materials or supported local employment?

These types of items are worth noting and worth talking about in terms of how they help you accomplish your mission.

Remember that there might be more organizations or publications interested in this type of achievement than your usual press list. Think of nonprofits, sustainability magazines or newsletters, and other local or regional water organizations. Especially if the achievement is given by another entity, make sure the word gets out that you are (humbly) taking steps to become more sustainable.

Community Workshops

One way to help members of your community learn more about sustainable products and behaviors is to sponsor community workshops. You may choose to offer these just to your customers or collaborate with other agencies to offer them to a wider audience. Anything that reduces resource usage or reduces waste is a good place to start. Don't overlap other organizations that are providing information on such issues as energy efficiency or vehicle idling unless it pertains specifically to the water industry. Present what you know best. Your water conservation program is the best place to start.

If there are rebates on specific items that require a little education, open an evening to those interested in water-efficient toilets and washers and water-efficient irrigation systems. An open house format works well for this, as you can invite various experts to come to staff stations showcasing and describing different products and answering customer questions one-on-one. If you expect a lot of attendees, allowing each expert 10 minutes to explain the product to everyone will probably be more efficient. Then those with additional questions can ask those later as they visit each station.

Workshops are a good tool, but need to be coupled with other tools to obtain behavior change. They are a good way to introduce ideas. At our district they rank low on ways people want to receive information and other reports confirm this. One obstacle is they ask the potential audience to commit time and energy to show up, so a lot of folks attend, but often they are people already predisposed to be enthusiastic about the issue of conservation or sustainability. Workshops rarely reach outside of this small niche market. They don't hit our masses, so we have to understand this when evaluating their value.

Obstacles aside, I would still recommend providing workshops. They can really disseminate information well when used in combination with other communications such as newsletters, Web site information, in-person assessments and rebates.

Jim Meierotto, Tualatin Valley Water District

Facilitating workshop opportunities allows you to promote various water-use efficiency measures and messages to specific types of customers, such as

The conservation department at Tualatin Valley Water District has offered informational workshops to their customers since 2006. Hosting workshops provides TVWD the opportunity to provide key information to specific customers with practical results.

For instance, we provide a one-hour course on the proper installation of soil moisture sensors for landscape professionals. Partnerships are key to effective workshops. Local, large, private businesses often have educational programs, and tapping into relationships with these businesses can uncover technical knowledge and teachers.

Corporations and private businesses are critical to the success of workshops, as they show buy-in from a credible source other than the government. Let these speakers know they are not there to sell their product or incorporate sales pitches, but rather to talk about the technology and industry. We usually let them wear their logoed shirts and bring brochures and business cards, as they need to be able to justify their time. We understand this and want to promote their opportunity to network.

The workshop can often actually become a networking hub that promotes more conservation projects. Make sure to involve a variety of businesses to offer a mix of viewpoints. Discover the key industry business stakeholders in your region and include everyone you can. You definitely don't want to favor one business over another. It has worked well here, as there are a couple of main players in irrigation and we have very strong relationships with them. You also can include speakers from universities and nonprofits as well as other researchers, scientists, and specialists.

Contributed by Jim Meierotto, Tualatin Valley Water District

residential homeowners, commercial property managers, facility engineers, and landscape construction and maintenance professionals. Topics might include:

- Cooling tower water treatment
- Water lean–grow green landscape irrigation
- Irrigation reduction and irrigation efficiency
- Irrigation Association's Certified Landscape Irrigation Auditor class
- Turf grass agronomics
- Property manager/landscape contractor's landscape irrigations
- Residential irrigation system maintenance
- Technical training on soil moisture sensors

Reporting

There are many good resources for developing sustainability reports, including the step-by-step guide to sustainability planning referenced in chapter 3. The size of your utility and the staff and financial resources you are able to devote to sustainability reporting will shape how you report your progress. You may choose to produce a separate sustainability report or to integrate sustainability into current reporting documents.

Performance Measures

If your utility uses performance measures, be sure to include objectives and targets that address such things as energy efficiency and community involvement among your measures.

Measures might include:

- Drinking water compliance rate (days in full compliance)
- Direct and indirect greenhouse gas emissions per year
- Gallons per capita per day (gpcd) of water used by single- and multi-family accounts
- Tons of hazardous waste disposed
- Operating cost coverage (total annual operational revenues/total annual operating costs)
- Debt service coverage ratio: Net revenues + fund balance/debt service
- Percentage of retail customers that rate the utility as good or excellent on customer satisfaction survey
- Nature and extent of environmental impacts attributable to the utility that disproportionately impact certain communities
- Number of people participating in apprenticeship, internship, and mentoring programs and percentage who go on to secure employment with the utility
- Average number of days lost through unscheduled absence (due to sickness)
- Percentage of employees with last performance review of 3.0 or better on 5-point scale (excluding retirements) that stay at the utility
- Average number of days to fill a vacant position
- Average number of formal training hours per employee, voluntary and mandatory
- Percentage of staff evaluating managerial leadership as good or better

Do not, however, just pick from the above measures without going through a process to determine what is truly vital to your utility's success. Do not just include sustainability-oriented measures just so you can say you have embedded sustainability in your performance measures. If your staff is not engaged in the process of figuring out what is important in your particular situation, the measures will not be tracked or used.

Take time with key staff to determine what measures are most important for you to follow on a regular basis. Visit www.staceybarr.com for some good information on how to do this. Then develop a reporting system so they are actually monitored.

Sustainability and Triple Bottom Line Reporting

A good sustainability plan can actually provide the basis for your reporting, though the detail with which you report may vary widely. From including a few paragraphs in your annual report, to producing a separate but basic sustainability report, to using the very detailed Global Reporting Initiative (GRI) structure, there are myriad ways to describe your efforts and results.

How to find the right measures is the most asked question in the field of performance measurement. And it's little wonder, because the more meaningful measures track outcomes that tend to be less tangible than the traditional things we've measured, like how many widgets we produced.

How do you translate results so intangible as employee morale, service quality, or corporate image into solid, robust measures?

The framework described here is an excerpt from the How-to Kit: How to Design Meaningful Performance Measures on staceybarr.com, which provides a systematic approach for taking almost all of the pain out of the challenge of finding the right measures.

- **Step 1: Begin with the end in mind.**

  Performance measures are objective comparisons that provide evidence of an important performance outcome. It is of the utmost importance to decide which outcomes are most worth tracking right now. As the first step in deciding how to measure an outcome, write down what the outcome is or what the difference is you are trying to create (and thus want to track using a measure). Focus on one outcome at a time.
- **Step 2: Be sensory specific.**

  When you have the end in mind, you are ready to get a handle on what specifically about your outcome you will measure. This is where you take care in your choice of words to describe the outcome as concretely as possible. Use "sensory" language–the language that describes what you and others would see, hear, feel, do, taste, or smell if your outcome were happening now. Avoid those inert words that we so often see in our goal and objective statements, such as efficient, effective, reliable, sustainable, and quality.
- **Step 3: Check the bigger picture.**

  Check the bigger picture for what could happen if you measure your outcome. What level of control do you have over achieving it? What might be the unintended consequences of measuring the outcome (both the positive and the negative)? What behavior would the measures drive? Which other areas of performance might be sabotaged or limited? This is your first chance to change your mind about what's most worth measuring.
- **Step 4: What's the evidence?**

  Now, get ultraspecific and figure out what the potential measures are that could let you (and everyone else) know that the outcome is being achieved. For each of your sensory-rich statements from step 2, what could you count on to tell you the extent to which it is occurring? Which of these potential measures would be the optimal balance between objectivity and feasibility?
- **Step 5: Name the measure.**

  Naming your performance measures marks the point at which you know exactly what you will be measuring. Be succinct, informative, and deliberate, as you need to be able to continually and easily identify each measure as it moves through the steps of being brought to life and being used in decision making.

Taking Action

Create your own measure design template based on these five steps. Now use your measure design template to start designing measures for the tricky goals, objectives, results, and outcomes you've struggled to measure thus far. Practice makes perfect!

Contributed by Stacey Barr (2009),
Queensland, Australia

Sydney Water's sustainability performance is reported annually in the Annual Report and is provided on Sydney Water's Web site at http://www.sydneywater.com.au/AnnualReport/SustainabilityIndicators/Sustainability_overview.cfm.

The reporting is structured under Sydney Water's corporate goals. Key performance areas that are critical to achieving these goals are identified in the organization's performance scorecard. The performance areas are evaluated based on Sydney Water's performance against its sustainability indicators, which show performance trends for 10 years of data where available. The performance scorecard captures Sydney Water's social, economic, and environmental sustainability performance: http://www.sydneywater.com.au/AnnualReport/AtAglance/Performance.cfm

Resources

Online Carbon Emissions Calculators

- www.safeclimate.net/calculator/
- www.b-e-f.org/carbon/calc

Performance Measures

- Walch, M., R. Immanni, S.W. Garrison, and J. Torbert. 2010. Improvement Through Performance Measurement. *Jour. AWWA*, 102:1:40.

Sustainable-living Calculators

- New Zealand's CarboNZero household calculator, www.carbonzero.co.nz
- Australian Conservation Foundation's Ecofootprint calculator, www.vic.greens.org.au/get-involved/sustainable-living
- London's Planet Positive calculator, www.planet-positive.org/calculator/
- Kid's calculator, www.planet-positive.org/how_2_kidscalc.php
- UK's Bioregional calculator, http://calculator.bioregional.com/
- US Geological Survey water usage calculator, http://ga.water.usgs.gov/edu/sq3.html

Sustainability Reporting Resources

- *American Institute of Certified Public Accountants (AICPA),* http://www.aicpa.org/innovation/baas/environ/faq.htm
- Corporate Register (compilation of sustainability reports), www.corporateregister.com
- Global Reporting Initiative (GRI), www.globalreporting.org/
- International corporate sustainability reporting, http://www.enviroreporting.com/

Each utility will have different needs in this area. You may report to elected officials, management, financial institutions, peers, customers, or the general public. As noted earlier in this chapter, each audience is looking for different material. Consider who will be reading the report and how much detail they will want in various areas. See Resources at the end of this chapter for more information.

REFERENCES

Barr, S. 2009. June 2, 2009. #23 *Five Steps to Find The Right Measures*. www.measure-up.blogspot.com. Accessed 1/19/10.

NWEI (Northwest Earth Institute). 2009. www.nwei.org. Accessed 12/6/09.

Walch, M., R. Immanni, S.W. Garrison, and J. Torbert. 2010. Improvement Through Performance Measurement. *Jour. AWWA*, 102:1:40.

Chapter 6

Where Do We Go From Here?

The water industry is currently tackling many issues, and elected officials, managers, and staff are being asked to face new challenges every day. I asked several water utility professionals to identify which issues were most pressing for water utilities. A selection of their responses includes:

- Dealing with changing demographics in the face of our ability to obtain additional sources.
- Energy usage and efficiency. Procedures that reduce our use of energy will serve us well in the future.
- More immediate issues (for example, energy efficiency, infrastructure replacement, source water protection, and supply development) are always important. Issues like water quality and conservation are "constants," which means that they get lost sometimes (for example, conservation is a lot easier to talk about during a drought than during a mild summer).
- Locally, I suspect water quality perception is the most challenging concern, largely because people don't have any real grasp on how good the water is or how onerous some of the other issues are.
- The combination of increasing regulation and increasing security leave challenges that may come into conflict with sustainability objects (for example, additional equipment, processes, and controls will likely increase use of energy and chemicals).
- Managing water demand versus supply.
- Climate change: while we don't know exact specific impacts (will sea level rise ½ in. or 2 ½ in. by 2050?), there are lots of things we can begin to do for mitigation, but perhaps more importantly, to begin the long-range capital planning for adaptation.
- Revenue and investment in infrastructure management and research and development (R&D).
- Sustainable procurement and supply-chain management.

- Balancing shared use and balancing social needs with environmental needs.
- Carbon, ecological, and social footprinting.
- Regulatory changes (including cap and trade) represent substantial unknowns in terms of future costs.
- We are highly vulnerable to sea level rise, both in terms of impacts to our water supply (saltwater intrusion) and in terms of inundation and tropical storms.
- More expensive alternative supplies may be required to meet future needs.
- Fostering culture change so that sustainability principles are enacted on the ground in an integrated manner at all levels, and in all business areas (i.e., going beyond the words so that all staff see sustainability as "doing business," not just a role for a few).

And what is cutting edge for water utilities? What will be the focus for the future for water utility sustainability? I heard:

- Generating our own energy.
- Recycled water—becoming more palatable as more drought years string together—and decentralized on-site water recycling.
- Figuring out different ways to manage the water supply will be the focus. For example, through a SCADA system, having a better handle on moving water to meet needs more efficiently.
- Better asset management. It leads to smart maintenance, more efficient operations and maintenance (O&M), and much better capital planning.
- Corporate reporting, that is, sustainability reporting without the need to define it as "sustainability reporting."
- Conducting sustainability assessments prior to capital investments/capital expenditures.
- Total water cycle management.
- Knowledge transfer/workforce planning.
- Teleworking/telecommuting.
- A mind set change from short-termism to long-term (whole of life) strategic planning.
- Working with communities to recognize that water is a precious and finite resource.
- The cutting edge is how to manage risk. One could be in the position to commit incredible amounts of money to climate change based on limited controversial data (i.e., is it a synthetic or natural cycle) in the face of a population that may or may not exist when the project is finished.
- Sustainability in the water sector will necessarily be a combination of conservation/demand-side management, new technology to meet the quality and distribution requirements of utilities, and more realistic pricing that reflects the actual costs of developing and maintaining more sustainable systems.

- Dramatic reductions in per capita water demand: US per capita demand is twice what it is in most other developed countries. We can continue to achieve significant energy savings and water resource savings. We just need to plan for it and encourage it. Savings in pumping, production, and capital investment would be staggering. But we'll cripple utilities' bottom lines if we can't plan effectively to encourage and fully realize these potential savings.

Our industry faces a broad range of issues—and many of them need to be our highest priorities. So how do we fit sustainability into the rest of what we do?

The biggest challenges in pursuing sustainable solutions?

- Finding the right software and ways to track what we want to track. Waste audits, for instance, are labor intensive, but that's the only way to show how our impact has changed. People never seem to have the time or know-how to track what we do.
- Maintaining water quality in light of being more energy efficient and conserving water. We do ozone treatment during high-demand periods and that is energy intensive. We do optimize so we only use ozone when we can't treat with other chemicals or processes.
- Getting people to understand that sustainability is more than just the environment.
- A short-term command and control mind-set.
- Time and money. We'd like to use hand-weeding and integrated pest management at all of our facilities, for instance, but we don't have enough people or money to make that happen.
- A methodology for valuing the triple bottom line and making each leg of the stool equitable.
- Competing interests—conservation versus revenue. The more successful a water utility is in driving conservation, the more rate revenue declines. Hence, there is a competing interest between conservation and the revenue needed for operational costs and capital projects.
- Promoting and sustaining a culture of sustainability. There is a tendency to think that sustainability is achieved or that it is someone else's responsibility. There is a constant temptation for focus on other challenges or for an organizational culture to revert to business-as-usual practices.
- Balancing different needs such as municipal, agricultural, wildlife, etc., with ever-increasing needs in all areas presents a challenge when planning with a limited resource.
- Aligning traditional "meeting-the-customer-needs" goals (which in our rapid growth environment, until recently, was a full-time job unto itself) with the less apparent and much less immediate goals associated with things like climate change. This is especially hard when there is resistance to the idea of increasing rates to make the investments in more sustainable equipment and practices.

- Understanding the full cost of the action/solution. Sometimes it requires something to be done differently, which can be a challenge. People focus on the short-term costs versus the long-term impacts and then deal with the fallout, if there is any. These are all costs.
- Water infrastructure is long-lived and expensive. You can't easily say "let's try it and if it doesn't work, we'll go back and do something different." Realizing significant changes will be challenging.
- Everything is connected to everything else: energy efficiency influences water quantity; conservation influences water quantity and revenue; water quantity and water quality are "joined at the hip"—you can't change one without impacting the other. It's hard to see the forest for the trees when it comes to sustainability (lots of opportunities to have unintended consequences), and it's really appropriate to plan accordingly and make baby steps in the right direction rather than assuming you can make a few big changes and achieve big (short-term) benefits.

Thousands of years ago, humans went out looking for water, finding ways to bring it to places where it was useful to their developing communities. The first approach toward water usage was one of expanding the availability of the resource. Supply management. When we needed more we found a new river, lake, or aquifer and constructed a way to get that water to where we were.

We have since moderated that approach to include conservation measures, and many utilities have moved toward a demand management philosophy, i.e., still needing new sources for growing demand, but encouraging conservation whenever possible and incentivizing reduced water consumption.

The next step will require a further shift in our thinking and many questions along the way.

- When we reach a level of water consumption that does not adversely affect the ecosystems we operate within, how do we just make do with that amount?
- Will it work to not look for new sources but concentrate on changing how we use water to significantly reduce what we need to take from nature?
- What processes do we use water for now that might be accomplished in some other way?
- Will we become the fueling stations of the future, providing the water on which vehicles run?
- Will we develop a system that pays large users of water to not use water during what we now call peak use times so we don't have to increase the size of our infrastructure?
- Is it practical to treat water in each neighborhood or subdivision to reduce the amount of potable water that is used where nonpotable water would be acceptable?
- Will we develop filtering systems as part of individual faucets that safely filter only the water that needs to be filtered, leaving irrigation, toilet flushing, and other water untreated?

Final Thoughts

Frankly, I don't know how things will pan out. I don't know how serious climate change and population growth will be over what period of time. I don't know how politics may skew any solutions developed by water professionals, environmentalists, or groups combining the two. I don't know what new technologies or tools will help us become more sustainable in our operations.

What I do know is that we can't count solely on technology to save the day. We can't just wait for someone else to require us to change our practices through regulation, and we don't need to wait. There are hundreds of practices, many listed in this book, that will start us on our way (or keep us moving) toward a new way of looking at water—as a service, not a product.

One of the ways this is described is the "soft path" for water, which takes its name from the energy soft path of the 1970s. Soft path water planning takes a step back from the problem and rethinks whether or not water is actually necessary to accomplish a particular task. If water is necessary, it then considers what quality of water is required to deliver the service desired and how that water can be used most efficiently (Brandes, 2005).

The soft path for water is a comprehensive management and planning approach that moves demand management to the next level. The soft path unleashes the full potential of demand management by simultaneously changing water-use habits, technologies, and practices. Working within ecological limits, the soft path promotes local public participation to ensure sustainability of our water resources. Soft path planning looks 20 to 50 years into the future and proposes major changes in our water infrastructure and institutions. The focus is on designing and implementing policies and strategies today that will meet new water demand through more efficient use of existing supplies, reducing or even eliminating the need for further supply-side developments (POLIS, 2009).

From the electrical power industry of the 1980s we get another model: integrated water resource management. Applied to water management, integrated resource planning is a systems approach that explores the cause-and-effect relationships affecting water resources wherever the planning entity's operations affect water use, quality, and supply. The process analyzes all the interrelated water management components in a given region. The focus is on the interrelation of the different water management components with the understanding that changes in the management of one component will affect the others. Because these components are often not confined to the boundaries of a single water management agency, a consensus-based, cross-jurisdictional, regional approach may be required to formulate comprehensive, win-win solutions to identified problems (California Water Plan Update, 2005).

Efficiency is about doing things a better way.
Conservation is about doing different things.
Brandes, Maas, and Reynolds, 2006

Core Principles of Soft Path Analysis

Treat water as a service rather than an end in itself. Beyond a very few services, there is no absolute requirement for water, or at least water beyond that which comes from rain. Rather, water is an input that can be delivered as an alternative to other ways of achieving the same result: air-based cooling, rain-fed agriculture, waterless sanitation.

Make ecological sustainability a fundamental criterion. Soft paths recognize ecosystems as legitimate users of fresh water and as the foundation of our economy. Environmental constraints are built in from the start to limit the amount of water withdrawn from natural sources and to establish conditions on the quality of water returned to nature.

Match the quality of water delivered to that needed by the end use. For both economic and physical reasons, it is almost as important to conserve the quality of water as to conserve its quantity. Soft path policies are designed from the start to match the quality of water supplied to the quality required by cascading water systems, ensuring that wastewater from one use becomes input for another use—from a washing machine to a garden, or from a cooling system to other industrial uses.

Plan from the future back to the present. Traditional planning starts from the present and projects forward to the future. Soft path planning does just the reverse through a technique called ***backcasting***. First, it defines a sustainable and desirable future state for society, at least as water sources and uses are concerned. It then works backward to identify policies and programs that will connect the future to the present.

This approach is especially commonplace in California, but utilities everywhere will benefit from the understanding that we must work together to preserve the water we have for myriad uses.

The days of always looking for new supply are over. We need to improve the productivity of the water we have and match its distribution to the needs of those who use it. And however we go about that, we must recognize that, in the long run, we do have to live within the limits of nature. I believe this is the next paradigm shift for water utilities.

Henry David Thoreau said, "Things do not change; we change."

Our future is up to us. Nature does not change, so we must adjust. We rethink our processes, we review our purchases, we refuse to believe that "the way we have always done it" is the best we can do. We can do better. And, working together, we will.

Resources

Soft Path for Water

- http://www.poliswaterproject.org/
- WaterSmart Scenario Builder—may help Canadian utilities in particular to integrate long-term water conservation scenarios into water supply master plans, http://www.poliswaterproject.org/toolkit#calculator

Listing of general resources on water conservation planning, especially in Canada

- Water Conservation Planning Quick Guide, http://www.poliswaterproject.org/publication/286

References

Brandes, O., and D.B. Brooks, Friends of the Earth, and POLIS Project on Ecological Governance. 2005. *The Soft Path for Water in a Nutshell.* Victoria, B.C.: University of Victoria.

Brandes, O., T. Maas, and E. Reynolds. 2006. *Thinking Beyond Pipes and Pumps: Top 10 Ways Communities Can Save Water and Money.* The POLIS Project on Ecological Governance. Victoria, B.C.: University of Victoria.

California Water Plan Update. 2005. Vol. 1, Chap. 2, Box 2-3. www.waterplan.water.ca.gov/docs/cwpu2005/vol1/v1ch02.pdf?/. Accessed 1/22/10.

POLIS Project on Ecological Governance Web site. 2009. http://www.poliswaterproject.org/softpath. Accessed 1/19/10.

Appendix A

Abbreviations

ASR	aquifer storage and recovery
CAA	Clean Air Act
CWA	Clean Water Act
EHS	environmental health and safety
EMS	environmental management system
EPR	enhanced product responsibility
FCA	full cost accounting
GHG	greenhouse gas
GRI	Global Reporting Initiative
HCS	hazard communication standard
IPM	integrated pest management
IRWMP	integrated regional water management plan
ISO	International Organization for Standardization
kW•h	kilowatt hour (of electricity)
LCA	life-cycle analysis
LCC	life-cycle costing
LCCA	life-cycle cost assessment
LEED	Leadership in Energy and Environmental Design
mgd	million gallons per day
MSDS	material safety data sheet
MW•h	megawatt hour (of electricity)
NWEI	Northwest Earth Institute

OHSAS	Occupational Health and Safety Assessment Series
OSHA	Occupational Safety and Health Administration
PBT	persistent bioaccumulative toxin
PCW	postconsumer waste
REC	renewable energy certificate or credit
RFP	request for proposal
SA	social accountability
SDWA	Safe Drinking Water Act
SHARP	Safety and Health Achievement Recognition Program
SMS	sustainability management system
TBL	triple bottom line
TCA	total cost analysis
TCM	total chemical management
TNS	The Natural Step
USEPA	US Environmental Protection Agency

Appendix B

Glossary

ABCD method	A strategic tool used in The Natural Step for backcasting from sustainability principles.
Asset management	Managing infrastructure capital assets to minimize the total cost of owning and operating them, while delivering the service levels customers desire. (Source: USEPA.)
Backcasting	A strategic planning method where a successful future is envisioned first. The current reality is then assessed against the vision, and a plan developed to bridge the gap.
Baseline assessment	Evaluation of performance in a baseline year that serves as a starting point for future evaluation of performance and long-term review of trends.
Basic human needs	A comprehensive set of fundamental human needs that are culturally and historically universal, nonoverlapping, nonsubstitutable, complimentary to one another, and must be satisfied on a continual basis. They are subsistence, protection, affection, idleness, identity, freedom, creativity, participation, and understanding. This list of basic human needs was created by the Chilean economist, Manfred Max-Neef, and is used to define human needs for the fourth system condition of The Natural Step. (Source: The Natural Step.)
Benchmarking	The process by which organizations evaluate their performance in relation to best practices within their own sector, or against their own past performance.
Best practice	An activity or procedure that has produced outstanding results in another situation (or organization) and could be adapted to improve effectiveness, efficiency, ecology, innovation, etc., in another situation (or organization).

Biodiesel	A clean-burning alternative fuel for compression-ignition (diesel) engines, produced from vegetable oils or animal fats.
Biodiversity	The variability among living organisms within species, between species, and between ecosystems. (Source: World Resources Institute.) Biodiversity is often used as a measure of the health of biological systems.
Biofuels	Fuels derived from plants (e.g., ethanol, biodiesel).
Bioretention cells	*See* Rain gardens
Biosphere	The portion of the Earth and its atmosphere capable of supporting life.
Capacity-building	The transfer of knowledge from experts to practitioners with the intent of strengthening the skills and ability necessary for them to act self-sufficiently.
Carbon offsets	Fees paid to offset carbon emissions through the greenhouse gas reduction actions of others.
Chemical life cycle	The entire spectrum of human health, environmental, and financial costs and liabilities surrounding the procurement, storage, use, and disposal of chemicals.
Cisterns and rain barrels	Cisterns and rain barrels harvest and store rainwater from roofs, which provides "soft" chemical-free water for garden or lawn irrigation, reduces water bills, and conserves municipal water supplies.
Clean Air Act (CAA)	The federal act that regulates air emissions from area, stationary, and mobile sources. CAA limits the emission of pollutants into the atmosphere in order to protect human health and the environment from the effects of airborne pollution.
Clean fuels	Clean fuels generate substantially fewer emissions than traditional diesel or gasoline. This may include natural gas, which does not qualify as a biofuel.
Clean Water Act (CWA)	The federal act that sets the basic structure for regulating discharges of pollutants to surface waters of the United States. CWA imposes contaminant limitations or guidelines for all discharges of wastewater into the nation's waterways.
Cloud computing	A way to deliver system solutions through the Internet drawing on a distributed or virtualized infrastructure. An organization using a cloud-based solution does not use its own server hardware, but draws on a shared infrastructure. This approach is more sustainable because cloud service providers are able to optimize and distribute computing resources across many organizations.

Cogeneration	The production of energy from a waste product, for example, through heat recovery. This term usually refers to electricity production, but the concept can be applied to biofuels (e.g., energy from cow manure in a dairy operation).
Commissioning	Process by which equipment, facility, or plant (which is installed, complete, or near completion) is tested to verify that it functions according to its design objectives or specifications.
Community engagement	The involvement of the community in the creation and implementation of major decisions.
Community planning	Any planning that is done for a community. This may include, but is not limited to, official community plans, municipal planning strategies, and/or integrated community sustainability plans. Community planning may encompass a variety of aspects, such as planning for sustainability, water use, economic development, parks, and/or land use.
Continuous improvement	In the context of management systems, the continuous focus on improving an organization's performance.
Corporate responsibility	An ethics-based business attitude that dictates that business leaders consider values and ensure their companies are accountable for implementing and acting on socially, environmentally, and financially responsible policies and practices.
Cradle to cradle	Concept of designing manufacturing systems such that materials either are returned to the "technical cycle" where they are used again or safely returned to nature.
Disposal	The discharge, deposit, injection, dumping, spilling, leaking, or placing of any liquid, gaseous, or solid hazardous waste on or in the land or water.
Downstream solutions	Solutions that deal with the symptoms of larger systemic problems. These solutions are often developed in reaction to the obvious symptoms without examining the cause of the symptom.
Eco-labels	Third-party certification (i.e., audited by a disinterested party) that guarantees certain levels of environmental or sustainable performance.
Ecoroof	Roof made of natural, living material intended to reduce runoff, mitigate the urban heat island effect, insulate the roof, extend its life, and reduce energy costs.
Ecosystem	A dynamic complex of plant, animal, and microorganism communities and their nonliving environment interacting as a functional unit.

Ecosystem services	Products like clean air and drinking water and processes such as the decomposition of wastes that benefit humankind and are supplied by natural ecosystems.
Energy efficient	Energy-efficient products or facilities use less energy than typical products or facilities to provide a given energy service or level of activity.
Environment	An organization's natural and human surroundings.
Environmentally preferable	Products or services that have a lesser or reduced effect on human health and the environment when compared with competing products or services that serve the same purpose. This comparison may consider raw materials acquisition, production, manufacturing, packaging, distribution, reuse, operation, maintenance, or disposal of the product or service.
Environmentally preferred	Environmentally preferred or green products are products and services that have less effect on human health and the environment than competing products and services that serve the same purpose. This comparison may consider the source of raw materials, production, manufacturing, packaging, distribution, potential for reuse and recycling, operation, maintenance, or disposal of the product.
Environmental health and safety	The area of an organization responsible for making sure a company is aware of, and adheres to, all applicable environmental and safety-related compliance requirements. Often responsible for emergency response planning, training, drills, monitoring, reporting, and the mitigation of releases of hazardous materials to the environment, and maintaining material safety data sheets (MSDSs).
Environmental management system	A systematic method for improving environmental performance. *See* also ISO 14001.
Extended producer (or product) responsibility	*See* Product stewardship.
Full cost accounting	A method of accounting where the full costs and benefits of the decision are weighed (i.e., costs are estimated for the long-term impacts of the decision; hidden costs, externalities, overhead, and indirect costs are included).
Funnel metaphor	A metaphor, used in The Natural Step, that demonstrates the sustainability challenge that results from continuing along our current path.
Global Reporting Initiative	An international standard for sustainability reporting sponsored by CERES.

Greenhouse effect	The rise in temperature that the Earth experiences because of a buildup of certain gases in the atmosphere. Without these gases, Earth's average temperature would not be warm enough for humans to live. Some greenhouse gases are produced naturally, and others from the burning of fossil fuels, land clearing, emissions from landfills, and other sources, including the manufacture, use, and disposal of products.
Greenhouse gases	Gases in the atmosphere that reduce the loss of heat into space. Human-induced emissions of greenhouse gases are believed to be a main driver of current global temperature increases and increasingly erratic weather.
Green roofs	*See* Ecoroof.
Green streets	Green streets replace conventional catch basins, pipes, curbs, and detention facilities with vegetated swales, bioretention cells, and/or pervious pavement.
Hazardous chemical	As defined by the Hazard Communication Standard (HCS), a hazardous chemical is one that can cause a physical or a health hazard, as determined by the chemical manufacturer.
Hazardous waste	A waste with properties that make it dangerous or capable of having a harmful effect on human health and the environment.
Indicator	A measure used to determine, over time, performance of a defined process, function, or achievement of a defined outcome, used to assess whether or not an organization is moving toward its vision and goals.
Integrated pest management	An approach to agriculture pest management that maximizes use of biological controls and only uses pesticides as a last resort.
Integrated regional water management plan	A comprehensive planning document to encourage regional strategies for management of water resources.
International Organization for Standardization (ISO)	The world's largest developer and publisher of international standards, ISO is a network of the national standards institutes of 157 countries, coordinating a system of standards that ensure desirable characteristics of products and services such as quality, environmental friendliness, safety, reliability, efficiency, and interchangeability at an economical cost.
ISO 14001	An international standard for environmental management systems.
Leadership in Energy and Environmental Design (LEED)	A third-party certification program that promotes a whole-building approach to the design, construction, and operation of high-performance green buildings.

Life-cycle analysis	The comprehensive examination of a product's environmental and economic effects throughout its lifetime, including new material extraction, transportation, manufacturing, use, and disposal.
Life-cycle cost assessment	*See* Life-cycle costing.
Life-cycle costing	The comprehensive accounting of the total cost of ownership, including initial costs, energy and operational costs, longevity and efficacy of service, and disposal costs.
Management system	A system of interrelated elements, including responsibilities, authorities, relationships, activities, functions, processes, practices, procedures, and resources.
Materiality	Refers to the degree to which an issue could make a major difference to an organization's performance.
Material safety data sheet	A document explaining safe use and accident response for products containing chemicals with potential human health impacts.
Mental models	Deeply ingrained assumptions, generalizations, or pictures and images that influence how we understand the world and take action.
Nontoxic CEO	A company leader who reduces where possible and responsibly manages at all times hazardous chemicals for the greater protection of people, the planet, and profits.
Normalization	A means of standardizing data to a particular measure of activity (e.g., million gallons per day of water delivered), such that true performance can be seen over time independent of variance in activity levels.
Occupational Safety and Health Administration	The US federal agency responsible for preventing work-related injuries, illnesses, and deaths.
OHSAS 18000	International standards for health and safety management systems. *See* http://www.ohsas-18001-occupational-health-and-safety.com/.
Permeable or porous pavements	Pervious, permeable, or porous pavements allow water to soak through the paved surface into the ground. Examples are porous concrete and asphalt, grasscrete, gravel with plastic grid systems, and interlocking paving bricks.
Persistent bioaccumulative toxins	A set of chemicals that are toxic, persist in the environment, and biologically accumulate in food chains.
Point-source discharges	Discharges of treated wastewater directly into a lake, river, stream, or other water body. Point-source discharges are regulated under the CWA.

Pollutants (contaminants)	Any element, substance, compound, or mixture that, after release into the environment and upon exposure, ingestion, inhalation, or assimilation into any organism, will or may reasonably be anticipated to cause illness, death, or deformation of the organism.
Pollution prevention	Avoiding, reducing, or controlling the creation, emission, or discharge of contaminants or waste materials.
Postconsumer	Postconsumer material generated by households or by commercial, industrial, and institutional facilities as end users of a product, which can then no longer be used for its intended purpose. This includes returns of material from the distribution chain.
Postconsumer waste (PCW)	Waste (generally paper) that has been used for its initial purpose, sent through the recycling system, and remanufactured into a new product. It does not include trimmings, etc., that never leave the mill.
Practicable	Practicable means the product is satisfactory in performance and available at a reasonable cost.
Precautionary principle	A framework that guides decision makers to take anticipatory and protective measures when an activity raises threats of harm to human health or the environment, even if some cause-and-effect relationships are not fully established scientifically.
Preconsumer material	Material diverted from the waste stream during the manufacturing process.
Procurement guidelines (aligned with sustainability principles)	A set of guidelines to help ensure that an organization's procurement decisions (including purchasing, inventory control, traffic and transportation, receiving, inspection, storekeeping, and salvage and disposal operations) help move society toward sustainability.
Product stewardship	A strategy designed to promote the integration of environmental costs associated with goods throughout their life cycles into the market price of the products, thereby sharing responsibility for the lifetime impacts of products. Also known as extended producer responsibility or extended product responsibility (EPR).
Rain gardens (bioretention cells)	Vegetated areas that collect runoff so it can slowly infiltrate into the ground. Some have special soil mixtures that maximize infiltration and pollutant removal but avoid extended ponding.
Recycled content products	Products made from materials that have been recovered, processed, and used as a raw material for the manufacture of a useful new product through a commercial process. These products will contain a specified percentage of material that would otherwise have been disposed of as waste.

Recycled materials	Components for incorporation into a product that have been reprocessed from recovered (reclaimed) material by means of a manufacturing process and made into another final product.
Remanufactured	To renew or restore a used product into its original form or into a useful new product through a commercial process.
Renewable energy (or renewables)	For electricity, this may include wind, solar, and low-impact hydro. For transportation, this may include biofuels and hydrogen. For heating, it may include solar thermal systems or methane recovered from biological processes (e.g., anaerobic digestion systems for manure or methane captured from landfills).
Renewable energy certificate	A tradable, nontangible energy commodity in the United States that represents proof that 1 megawatt-hour (MW•h) of electricity was generated from an eligible renewable energy resource.
Resource Conservation and Recovery Act (RCRA)	The act that establishes a framework for national programs to achieve environmentally sound management of both hazardous and nonhazardous wastes and promotes resource recovery techniques and methods to reduce the generation of hazardous waste.
Risk	The probability of a consequence occurring multiplied by the magnitude of the consequence.
Root cause	The most basic reason for the presence of a problem, which, if eliminated, would prevent its recurrence.
SA 8000	A verification system for ensuring humane workplaces.
Safe Drinking Water Act (SDWA)	The act designed to protect the nation's drinking water supply by establishing national drinking water standards.
Smart site design	Design that conserves trees and habitat, minimizes disturbance and soil compaction, and integrates on-site stormwater management into the other considerations for site development.
Source reduction	Maximizing or reducing the use of natural resources at the beginning of an industrial process, thereby eliminating the amount of waste produced by the process. Source reduction is the preferred method of waste management.
Stakeholder	Any person or group who has an interest (a stake) in an organization or community.
Strategic planning	The process of identifying an organization's long-term goals, determining the best approach for achieving those goals, and allocating adequate resources (including capital and people) for implementation.

Sustainability	A means of configuring civilization and human activity so that society, its members, and its economies are able to meet their needs and express their greatest potential in the present, while preserving biodiversity and natural ecosystems, and planning and acting for the ability to maintain these ideals in the very long term. (Source: Wiktionary.org.)
Sustainability framework	A mental model for understanding what sustainability means. There are numerous frameworks that include the triple bottom line (economic, social, and environmental), The Natural Step, and the Earth Charter.
Sustainability management system	A systematic method for improving sustainability performance (similar to EMS but includes more than just environmental criteria).
Sustainability plan	A plan that identifies key priorities, actions, targets, responsibilities, resources, and timetables for a business or organization to implement its environmental, economic, and social strategies for improving overall performance.
Sustainability reporting	The act of publicly reporting on environmental, social, and economic management and performance. The internationally recognized standard is the Global Reporting Initiative's Sustainability Reporting Guidelines.
Sustainable development	Defined in the 1987 Brundtland Report (Our Common Future) as development that "meets the needs of the present without compromising the ability of future generations to meet their own needs."
Sustainable procurement	The purchase of materials, products, and services in a manner that integrates fiscal responsibility, social equity, and community and environmental stewardship.
System conditions	Four conditions for achieving sustainability. The conditions are stated in the negative to create the constraints within which creativity can flourish. In a sustainable society, nature is not subject to systematically increasing concentrations of substances extracted from the earth's crust, concentrations of substances produced by society, degradation by physical means, and, in that society, people are not subject to conditions that systematically undermine their capacity to meet their needs.
Systems thinking	An approach to problem-solving that assumes that the individual problem is part of a much larger system. The intent is to solve the problem in a way that does not create further problems down the road. This approach is particularly important in complex systems where we do not always understand the interconnection between parts.
Technosphere	Refers to the human system as a subsystem within the biosphere, in which materials, industries, and their products interrelate and interact.

The Natural Step (Organization)	An international not-for-profit organization founded in Sweden in 1989 by Swedish scientist Karl-Henrik Robèrt. The Natural Step has pioneered a "Backcasting from Principles" approach to effectively move society toward sustainability. The organization is committed to helping to create a sustainable human society using The Natural Step Framework to communicate the scope and scale of the current problem and empower individuals to move toward strategic solutions.
The Natural Step Framework	A planning and decision-making framework that allows individuals to understand the root causes of unsustainability and then move strategically toward sustainability.
Total chemical management	The conscious and responsible corporate management of chemicals by way of life-cycle costing and throughout the chemical life cycle.
Toxicity	The degree to which a substance or mixture of substances can harm humans or animals.
Toxic substance	A chemical or mixture that may present an unreasonable risk of injury to health or the environment.
Toxic waste	A waste that can produce injury if inhaled, swallowed, or absorbed through the skin.
Triple bottom line	A way to evaluate organizational performance that takes into account the economic, environmental, and social impacts associated with the organization's activities.
Triple bottom line reporting	Systematic management, performance, improvement, and public disclosure in environmental, economic, and social dimensions at local, regional, and global scales. (Source: AwwaRF, 2007.)
Upstream solutions	Proactive solutions that address the source of the original problem, as opposed to the effects of it.
US Environmental Protection Agency	The US federal agency responsible for protecting human health and safeguarding the natural environment—defined as the air, water, and land.
Vegetated swales	Broad, shallow channels that reduce stormwater volume and velocity and filter pollutants from the water.
Xeriscaping	Landscaping that requires little or no additional watering.
Zero waste	The reuse or recycling of all materials back into nature or the marketplace in a manner that protects human health and the environment.

Appendix C

SCORE Assessment—Water Sector

WATER UTILITIES			
Practice	1 point Incubator	3 points Initiative	9 points Integrated
	Management and Policy		
Water Conservation: Stabilize absolute water usage at a sustainable level.	Provide water conservation program that educates customers about important water conservation issues and incentivizes them, through rebates and other means, to reduce consumption.	Manage by reducing demand and making current practices more efficient. Develop integrated watershed management plan with other entities in your watershed. Document a reduction in absolute peak demand.	Operate with best practice use of "fit for purpose" water from a sustainable supply that protects natural systems within the watersheds supplying the system. Set absolute system water usage limit and establish a plan that meets that supply level sustainably (i.e., without lowering groundwater level). Review supply limit and monitor environmental impact at least biennially.
Watershed Protection: Preserve watershed health.	Actively participate in local groups looking at water issues from a watershed perspective.	Promote public policy and practices that preserve the water quality and health of your watershed.	Ensure that policies and practices in your watershed(s) protect the availability and quality of water for humans and the ecosystem in perpetuity (e.g., water rights, land use/development, forest/agricultural practices, pollution prevention, etc.).
Sustainability Planning and Thinking: Increase sustainable planning mind-set.	Provide education and communication materials on sustainable thinking to every department.	Manage planning process by requiring specific performance criteria in all planning documents. Develop a "mind-set shifting campaign" to transition planners from traditional thinking to sustainable thinking.	Enable an organization that embeds sustainable thinking in every facet of training, operations, performance, and accountability.

Engineering and Field Operations			
Life-Cycle Thinking: Apply life-cycle thinking to your products and processes.	At least every two years, conduct life-cycle costing (LCC) on a major product used in your operations, e.g., a type of pipe or a category of vehicle.	Incorporate LCC into capital improvement plan, evaluating projects over their whole life from a triple bottom line perspective and implementing those with lower LCCs.	Design water systems that minimize moving water around, maintain quality, and recharge the natural systems within the watersheds supplying the system.
Emergency Preparedness: Have effective plans in place to protect citizens, property, and the environment in the case of natural or man-made disasters.	Regularly participate in community or regional exercises revolving around potential threats and how to provide an adequate supply of quality water in an emergency.	Have an effective network of trained staff members and a well-tested plan for foreseeable disasters.	Have systems in place for providing potable water without electricity or other major infrastructure elements.
Energy: Reduce environmental and social impacts associated with energy.	At least every five years, conduct an energy audit and process optimization audit, including a review of equipment startup and process sequencing to identify opportunities to reduce peak loads. Act on the results.	Have in place systems for monitoring and reducing impacts from equipment, processes, and human behavior. Identify energy generation opportunities in the supply and distribution systems (in-line hydroelectric for large head drops, pressure-reducing valves). Purchase or produce at least 25% renewable energy.	Purchase or produce at least 80% renewable energy (electricity and other fuels). Demonstrate significant overall reductions in energy consumption.
Waste Management in Field Operations: Eliminate the concept of waste.	Conduct a waste audit. Separate waste from recyclables in construction and demolition debris. Achieve 50% (by weight) average waste diversion from landfill (by weight).	Have a formal waste recycling policy that focuses on the avoidance of waste creation. Recycle or reuse at least 75% of any waste that is created on job sites.	Achieve "zero-waste"—minimize waste creation and average at least a 95% recycling/reuse rate for all construction and maintenance projects.
Stormwater Management: Manage stormwater related to construction.	Keep all sedimentation on site and out of the water system.	Have a formal stormwater management policy for utility projects and continue to measure and report on stormwater management.	Evaluate project's potential stormwater impact and implement appropriate measures to dispose of runoff by means consistent with the assimilative capacity of the site.

Appendix D

Energy and Greenhouse Gas Management Options in the Water Utility Sector

Contributed by
Steve A. Conrad
Simon Fraser University
School of Resource and Environmental Management
Burnaby, B.C., Canada

For water utilities, energy considerations represent a major concern that must be addressed if they are to continue to deliver sustained levels of service to customers. Utilities are currently being challenged in respect to providing water services by a range of factors including: population-driven escalation in demand; pressure on traditional water sources due to pollution, overextraction, climate variability, and competing demand from other sectors; more stringent environmental regulations; and volatility in energy supply and prices.

Whereas many utilities exercise strict control over water supplies, water quality, water treatment, and water operations, few manage energy as rigorously despite significant energy consumption and costs. Urban water systems require substantial energy inputs, and most new water supply solutions and treatment technologies involve additional energy inputs (Kenway et al., 2008). Energy is required for all stages of the urban water cycle, from the collection, pumping, and treatment of drinking water, through the transmission and distribution of supply to customers and the collection and treatment of wastewater (AwwaRF, 2007). Energy use in the water and wastewater sector will grow by 40% over the next 20 years, driven by new regulations and population growth (Arpke and Hutzler, 2006). Moreover, the anticipated introduction of energy-intensive technologies, such as desalination, will further increase energy use.

The water and wastewater utility sector makes a relatively small contribution to total national energy use and greenhouse gas (GHG) emissions compared with other sectors. In the United States, 3% of total electrical power generation, or

75 billion kW•h per year, is consumed by publicly owned water and wastewater utilities (Cohen et al., 2004). Yet, for utilities, energy consumption represents a very large portion of total operating costs, and their total energy consumption ranks among the highest in a great majority of communities. For instance, energy use represents a significant portion of total operating costs in the California State Water Project, accounting for approximately one third of the $900 million operating budget (Cohen et al., 2004).

This paper describes various aspects of energy use within water utilities and examines options for managing and reducing this use. Because utilities face significant difficulty managing energy across their entire water systems, this paper concludes by discussing barriers to implementing energy management decisions.

Energy Management

In the case of utilities, most energy management strategies focus on direct energy consumption of water and wastewater treatment and pumping facilities but omit water end use (e.g., residential, industrial, agricultural, and commercial use). The management focus regarding GHG emissions is similar. Kenway et al. (2008) note that total emissions through the full water life cycle are often overlooked, focusing solely on direct energy use emissions.

Many utility operations consume energy; consequently, energy management could potentially be applied in a large number of areas. Discussions about energy management options tend to focus on how to allocate water and protect drinking water supplies while minimizing energy consumption. Utilities balance competing goals in trying to achieve increasingly stringent water quality standards, while at the same time aiming to reduce energy consumption. However, factors including climate change, uncertainty of supply, and recent GHG regulations have prompted utility managers to consider new options (Proust et al., 2007; IPCC, 2007; Dessai and Hulme, 2007).

The Water Services Association of Australia (WSAA) 2007 report *Energy and Greenhouse Mitigation Strategies* has identified steps utilities are taking to reduce energy use and GHG emissions that include:

- avoiding energy use where other options exist to achieve the same outcome;
- improving energy efficiency measures (e.g., deploying advanced controls for pump optimization, implementing energy management systems, and installing variable speed pumps in place of fixed speed pumps);
- using wastes to generate renewable power; and
- sequestering carbon and purchasing carbon offsets.

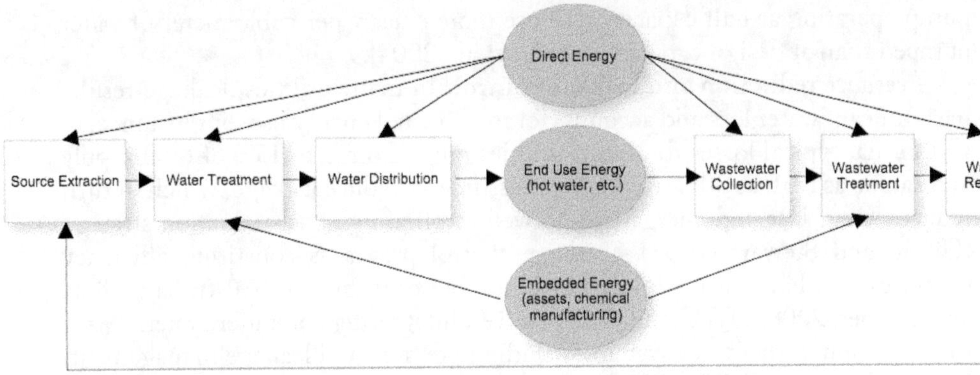

Figure D.1. Energy consumption in an urban water system

Energy Consumption in the Urban Water Cycle

The relationship between energy use and water delivery is mutually reinforcing, and utilities benefit from closely examining energy use internally and energy use driven by water demand. Figure D.1 illustrates energy consumption in a typical urban water system. Although end use is included here, there is no consensus on whether or not utilities should take responsibility for managing end use energy consumption.

Energy Management Options

Pumping. Depending on topology and distance, pumping represents significant energy use for many water and wastewater systems. Cities like Vancouver, B.C., have lower pumping energy requirements as they rely on gravity for distribution from pristine sources. In contrast, pumping energy increases in areas using only groundwater systems (Elliott et al., 2003). For instance, in cities like Tucson, Ariz., pumping energy requirements are relatively great even when compared to those for water treatment. In Wisconsin, pumping water accounts for 85% of the energy required to provide drinking water (Kenway et al., 2008). In addition, utilities with complex transmission systems, such as those in Northern and Southern California, face considerable pumping requirements. Pumping water over the Tehachapi Mountains, the highest point of any water system in the world (Cohen et al., 2004), to Southern California requires significant energy, comparable only to desalination (Stokes and Horvath, 2006).

Increasing the efficiency of pumping operations offers significant scope for reducing energy use. Technical approaches include installing efficient pumps and variable speed drives, and using more effective instrumentation and controls to manage pumping operations. However, developing or implementing universal approaches to increasing energy efficiency is difficult given that energy use is not always in a linear relationship to the volume of water transported or used. For example, pumps are more efficient when operating at design capacity, so a pump operating at half capacity will use more energy per cubic meter of water pumped than at design capacity (Cohen et al., 2004).

Pressure reduction and leakage control. In many utilities, leakage results in significant water loss and accounts for misdirected energy use. For instance, in California, typical losses for urban utilities range from 5 to 15% of total supply and can be as high as 30%. Reducing pressure can reduce leakage, which in turn reduces water loss and energy use, as well as offering a cost-reduction strategy (Girard and Stewart, 2007). Leakage-control programs constitute efficiency measures that have been explored by a number of utilities in Australia (Schott and Cooper, 2008; QWC, 2008). Moreover, integrating computer systems with the operation of the water system, including pressure and leakage management, has saved Fresno, Calif., $725,000 annually in energy and water costs (Cohen et al., 2004).

New water technologies. Adopting new water supply technologies will affect a utility's overall energy requirements. New water technologies such as desalinization use large amounts of energy per unit of water produced (Cohen et al., 2004). The amount of energy new technologies use depends on the quality of the input water and the process (Meerganz von Medeazza and Moreau, 2007). However, it is expected that new technologies will likely increase a utility's energy footprint (Proust et al., 2007).

With the introduction of new water quality technologies, existing supplies may also require additional energy (Stokes and Horvath, 2006). There is a trend in urban water services for more energy-intensive treatment methods aimed at disinfecting supplies, for example, ozonation and ultraviolet radiation (Cohen et al., 2004; Elliott et al., 2003).

Energy requirements to prepare water recycling as a source are a function of the treatment method needed to treat wastewater to quality levels required by end users and the energy required to convey the treated wastewater to the point of use (Cohen et al., 2004). Identifying the level of water quality required provides the opportunity to substitute lower quality water for high-quality potable water.

New wastewater technologies. Future trends in wastewater management, such as new combined sewer overflow regulations, watershed recovery programs, and emission regulations, will significantly influence energy use in wastewater treatment plants (NYSERDA, 2008). New technologies for addressing these trends will challenge utilities trying to achieve increasingly stringent wastewater discharge limits, while at the same time reducing energy consumption (Gaterall et al., 2005). Improving discharge performance will in most cases incur the cost of more energy-intensive treatment operations. Utilities may need to consider all environmental costs associated with different wastewater treatment processes in order to evaluate the most energy-efficient ones (Karlsson, 1996).

Approaches to reducing operational energy use in wastewater treatment processes include using renewable energy sources or energy from waste sources and developing high-efficiency mechanical and electrical plants. One potential alternative is the application of pretreatment technologies with low energy requirements capable of reducing the organic load entering the secondary treatment stage. An example of this is anaerobic reactors (Gaterall et al., 2005; NYSERDA, 2008).

Decentralized systems. Switching to distributed water supplies could considerably reduce energy requirements. For instance, widespread installation of rainwater tanks aimed at supplementing water supplies for domestic laundry, toilet, and outdoor uses could reduce operating costs and greenhouse gas emissions for regional water utilities or wholesale water suppliers (Coombes, 2006). However, to achieve cost and energy savings in this fashion, there are a number of factors to consider in the design of such systems and their interaction with existing water supplies (Marsden Jacob Associates, 2007).

Demand management and efficient end use. With respect to the provision of water and wastewater services, demand management and end-use efficiency measures, for example, water-efficient showerheads and gas or solar hot water systems, can significantly reduce energy use and GHG emissions. In the United States, 93 to 97% of the energy consumed in the domestic water cycle was for hot water heating, and it was estimated that in 1990 over $15 billion was spent heating residential water.

de Monsabert et al. (1998) estimate that in the United States the breakdown with respect to energy sources for residential water heating is 35% electric, 60% natural gas, and 5% others, such as fuel oil and solar. In Australia, GHG emissions associated with operating household appliances using water comprised between 90 and 97% of total system-wide annual GHG emissions (Flower et al., 2007). Moreover, in Australia, water heating accounts for approximately 85% of the combined system-wide energy requirements for both the water and wastewater sectors. In addition, it is estimated that every time a resident runs a hot water tap to expel cooled water in the pipes, approximately 5.06 m^3 (1,336.7 gal) of the previously heated water is lost (Cheng, 2002).

Upgrading domestic water heaters from electricity to either gas or solar can offset all the energy consumed by water and wastewater utilities. Furthermore, water-efficient showerheads can save both water and energy (Kenway et al., 2008), and the use of natural gas for heating water instead of electricity would significantly reduce the energy currently consumed for heating water (Arpke and Hutzler, 2006). Water utilities, such as Sydney Water, have introduced programs aimed at installing water-efficient showerheads, earning carbon credits as a consequence (Schott and Cooper, 2008). Initiatives like these could offset all carbon dioxide emissions if implemented across the customer base (Lundie et al., 2004).

Energy recovery. Utilities can offset much of their energy use by generating power using methane recovered from waste processes (Karlsson, 1996; Zhang and Wilson, 2000; NYSERDA, 2008) or by using hydraulic head to generate power (Kenway et al., 2008). For instance, utilities such as Sydney Water in Australia currently have three large cogeneration plants and another five approved for construction (Schott and Cooper, 2008). Sydney Water also has a number of hydroelectric plants approved for construction.

Life-cycle inventories. Utilities are beginning to develop life-cycle inventories focusing on energy use and GHG emissions. For instance, the City of

Toronto developed inventories for three processes within its water treatment plants. Inventories were based on chemical production, transportation of materials, and water treatment plant operations (Racoviceanu et al., 2007). Following the initial assessment, operations accounted for 94% of total energy use and 90% of GHG emissions.

Operational energy decisions. Operations at water utilities are typically geared to meeting consumption levels and water pressure (Jentgen et al., 2007), i.e., wells and booster pumps are automatically controlled based on reservoir levels and distribution pressures. Energy and cost savings are possible through the use of advanced control systems designed to optimize energy use, minimize cost, and meet water quality standards. Energy management frameworks, such as the Energy Water Quality Management System (EWQMS), provide a means to capture these benefits. EWQMS is a collection of individual application software programs that provide information to decision makers used to solve water quality and energy management problems associated with achieving the company's mission (Jentgen et al., 2005).

The application of software in calculating the optimal pumping schedule has been shown to reduce the operating cost of a large-scale water network by 14% while still satisfying all operating criterion (Bounds et al., 2006). Moreover, in San Diego, application of EWQMS resulted in a 15% reduction in energy consumption (Conrad and Cook, 2005). The return on investment for implementing advanced operational decision support systems is large, and the payback can be achieved in less than five years (Jentgen et al., 2005).

Other Energy Management Options

The preceding section described some of the energy management options that have been adopted by water and wastewater utilities. This does not represent, however, an all-inclusive list of available options. Energy management innovation continually evolves, and the options described in this report represent only a small portion of those currently under consideration by utilities. The following lists some additional options:

Management options:

- Energy audits
- Energy and water quality management systems (EWQMS) (control technologies)
- Asset management (full life-cycle analysis)
- Energy forecasting and demand response
- Energy partnerships
- Energy rate optimization

Facility improvements and field operations:

- Lighting and heating, ventilation, and air conditioning (HVAC) improvements
- Use of fuel-efficient fleet vehicles

Water distribution:

- Modeling and pressure management
- Postflocculation in the distribution system, water mains, and storage facilities
- Distribution system piping
- Pumping system evaluation
- Pump size selection
- Pump optimization
- Maintenance and repair programs
- Variable frequency drives
- Pressure-reducing valves and in-line turbines (micro-hydroelectric generation)
- Optimize gravity flow
- Premium efficiency motors
- Storage level optimization

Water treatment:

- Optimize operations
- River bank filtration
- Conventional filtration treatment
- Direct filtration treatment
- Diatomaceous earth filtration
- Hydraulic flocculation
- Polyaluminum chloride
- Air stripping
- Membranes choice
- Ozone optimization
- Ultraviolet disinfection optimization

Water management:

- Water loss audits
- Supply-side management
- Leak reduction
- Metering and supply programs
- Water use planning
- Alternative supply
- Demand-side management
- Water-efficient device incentive programs
- Metering and customer behavior programs
- Optimize irrigation
- Commercial and industrial water efficiency
- Education and outreach
- Conservation rate structures

Alternative/renewable energy sources:

- Solar power
- Wind turbines

- Geothermal
- Lake/ocean water cooling
- Micro-hydro power generation

Waste to energy:

- Lake water cooling

Energy Management Barriers

Opportunities for implementing energy management are numerous. However, many barriers remain, requiring energy managers to pay special attention to implementation approaches that will ensure the ability to make decisions and, once made, that the decisions will be implemented. This section briefly discusses some of the barriers to energy and GHG management.

Barriers inhibit effective energy management decisions throughout the utility organization. The following identifies those that have been most discussed in the literature:

> Operational barriers: Organizational staff typically has distinct roles in facility operations and crossover responsibility is limited. This segmentation of staff roles and responsibilities reduces the ability to implement broad energy management programs (NYSERDA, 2008).
>
> Institutional barriers: Water utilities are responsible for the provisioning of the urban water cycle and do not see themselves in the business of energy management. This creates difficulty in accessing the energy data required for energy accounting and changing operations to maximize energy efficiency and reduce costs (Jentgen et al., 2005).
>
> Regulatory barriers: Energy is not often regulated within water utilities, and compliance with water regulations often conflicts with optimal energy and GHG management options. Water managers often select conservative options (such as over-sizing pumps) to ensure regulatory compliance, or face resistance from executive management when opting for new technologies without a previous proven life cycle (Arpke and Hutzler, 2006).
>
> Financial barriers: Most utilities operate under tight budget constraints and are either reluctant to increase those budgets or are unable to do so owing to regulatory constraints. Energy management is then tied to cost savings, not efficiency savings, creating difficulties for utilities that wish to pursue energy efficiency or GHG reductions that do not demonstrate cost savings (Golove and Eto, 1996).
>
> Information asymmetries: Information is often held within a single group and not made available to managers to assist them in making decisions (Golove and Eto, 1996). Lack of information makes it difficult to identify the full impact of a decision.
>
> Stakeholder barriers: As is the case with political barriers, utility managers must respond to stakeholder (e.g., customers, upstream/

downstream utilities, government agencies) concerns that often conflict with energy management. While stakeholders may hold conflicting views, management decisions are better supported when they participate in the decision-making process (Breuer et al., 2008).

Risk management barriers: Management concern over uncertainty is often stated as a reason for inaction (AwwaRF, 2008). While risk and uncertainty will always feature prominently in energy management, a firm understanding of risk and how utilities can address it will allow energy management under uncertainty (Salewicz and Nakayama, 2004).

CONCLUSION

As water and wastewater treatment processes become more energy-intensive, water utilities will face a significant challenge with respect to identifying the best strategic and operational energy management options for effectively managing and ultimately reducing energy consumption and greenhouse gas emissions. Utilities will immediately benefit from actively engaging in energy management. Further, the uncertainty around climate change and the importance of being able to develop robust adaptation strategies center on a utility's ability to manage its water resources while continuing to meet GHG commitments.

ACKNOWLEDGMENTS

Research contained in this paper was supported by the Water Research Foundation (Project 4090). Project 4090 aims to develop a decision support system for use by water and wastewater utilities that wish to achieve sustainable energy management. Additional information on project 4090 findings will be available in the project report to be published in late 2010.

REFERENCES

Arpke, A., and N. Hutzler. 2006. Domestic Water Use in the United States—A Lifecycle Approach. *Journal of Industrial Ecology,* 10:(1–2); 169–184.

AwwaRF. 2007. *Energy Index Development for Benchmarking Water and Wastewater Utilities.* Denver, Colo.: Water Research Foundation.

AwwaRF. 2008. *Risks and Benefits of Energy Management for Drinking Water Utilities.* Denver, Colo.: American Water Works Association Research Foundation.

Breuer, N.E., V.E. Cabrera, K.T. Ingram, K. Broad, P.E. Hildebrandt. 2008. AgClimate: A Case Study in Participatory Decision Support System Development. *Climatic Change,* 87:385–403.

Bounds, P., J. Kahler, and B. Ulanicki. 2006. Efficient Energy Management of a Large Scale Water Supply System. *Civil Engineering and Environmental Systems,* 23:(3); 209–220.

Cheng, C. 2002. Study of the Inter-relationship Between Water Use and Energy Conservation for a Building. *Energy and Buildings,* 34:(3); 261–266.

Cohen, R., B. Nelson, and G. Wolff. 2004. *Energy Down the Drain: The Hidden Costs of California's Water Supply.* New York: Natural Resources Defense Council.

Conrad, S., and W. Cooke. 2005. Meeting and Exceeding Performance Measures in San Diego Using Water System Optimization. Presented at AWWA Annual Conference, San Francisco.

Coombes, P.J. 2006. Energy and Economic Impacts of Rainwater Tanks in Urban Areas on the Operation of Regional Water Systems. 30th Hydrology and Water Resources Symposium. Launceston, Tasmania.

de Monsabert, S., and B.L. Liner. 1998. Integrated Energy and Water Conservation Modelling. *Journal of Energy Engineering,* 124:(1); 1–32.

Dessai, S., and M. Hulme. 2007. Assessing the Robustness of Adaptation Decisions to Climate Change Uncertainties: A Case Study on Water Resources Management in the East of England. *Global Environmental Change,* 17; 59–72.

Elliott, T., B. Zeier, I. Xagoraraki, and G.W. Harrington. 2003. *Energy Use at Wisconsin's Drinking Water Facilities.* Report 222–1. Madison, Wis.: Energy Center of Wisconsin.

Flower, D., V. Mitchell, and G. Codner. 2007. The Potential of Water Demand Management Strategies to Reduce the Greenhouse Gas Emissions Associated With Urban Water Systems. 1st Conference on Sustainable Urban Water Management and 9th Conference on Computing and Control in the Water Industry. Leichester, UK.

Gaterell, M., P. Griffin, and J. Lester. 2005. Evaluation of Environmental Burdens Associated With Sewage Treatment Processes Using Life Cycle Assessment Techniques. *Environmental Technology,* 26:(3); 231–249.

Girard, M., and R.A. Stewart. 2007. Implementation of Pressure and Leakage Management Strategies on the Gold Coast, Australia: Case Study. *Journal of Water Resources Planning and Management-ASCE,* (133); 210–217.

Golove, W.H., and J.H. Eto. 1996. *Market Barriers to Energy Efficiency: A Critical Reappraisal of the Rationale for Public Policies to Promote Energy Efficiency.* Berkeley, Calif.: Lawrence Berkeley National Laboratory.

IPCC. 2007. *Climate Change and Water.* Geneva, Switzerland: Intergovernmental Panel on Climate Change.

Jentgen, L., S. Conrad, H. Kidder, and T. Lee. 2005. *Optimizing Operations at JEA's Water System.* Denver, Colo.: Water Research Foundation.

Jentgen, L., H. Kidder, B. Hill, and S. Conrad. 2007. Energy Management Strategies Use Short-term Water Consumption Forecasting to Minimize Cost of Pumping Operations. *Jour. AWWA,* 99:(6); 86.

Kenway, S., et al. 2008. *Energy Use in the Provision and Consumption of Urban Water in Australia and New Zealand.* Melbourne: Water Services Association of Australia.

Karlsson, I. 1996. Environmental and Energy Efficiency of Different Sewage Treatment Processes. *Water Science and Technology,* 34:(3-4); 203–211.

Lundie, S., G. Peters, and P. Beavis. 2004. Life Cycle Assessment for Sustainable Metropolitan Water Systems Planning. *Environmental Science & Technology,* 38:(13); 3465–3473.

Marsden Jacob Associates. 2007. *The Cost-effectiveness of Rainwater Tanks in Urban Australia.*

Marsh, D.M., and D. Sharma. 2006. Energy-water nexus: a review of existing models. *1st Australian Young Professional Conference.* Canberra: Australian Government National Water Commission.

Meerganz von Medeazza, G., and V. Moreau. 2007. Modelling of Water-energy Systems. The Case of Desalination. *Energy,* 32:(6); 1024–1031.

NYSERDA. 2008. *Statewide Assessment of Energy Use by the Municipal Water and Wastewater Sector.* Albany: New York State Energy Research and Development Authority.

Proust, K., S. Dovers, B. Foran, W. Newell, and P. Troy. 2007. *Climate, Energy and Water: Accounting for the Links.* Canberra: Land & Water Australia.

QWC. 2008. *South East Queensland Water Strategy.* Queensland (Australia) Water Commission.

Racoviceanu, A., B. Karney, C. Kennedy, and A. Colombo. 2007. Life-Cycle Energy Use and Greenhouse Gas Emissions Inventory for Water Treatment Systems. *Journal of Infrastructure Systems,* 13:(4); 261–270.

Salewicz, K., and M. Nakayama. 2004. Development of a Web-based Decision Support System (DSS) for Managing Large International Rivers. *Global Environmental Change,* 14:Suppl 1; 25–37.

Schott, K., and D. Cooper. 2008. Climate Change and Carbon Neutrality—Sydney Water's Strategy. *AWA Accounting for Carbon in the Water Industry.* Sydney, Australia.

Stokes, J., and A. Horvath. 2006. Life Cycle Energy Assessment of Alternative Water Supply Systems. *International Journal of Life Cycle Assessment,* 11:(5); 335–343.

WSAA. 2007. *Energy and Greenhouse Mitigation Strategies.* Melbourne: Water Services Association of Australia.

Zhang, Z., and F. Wilson. 2000. Life-Cycle Assessment of a Sewage-Treatment Plant in South-East Asia. *Water and Environment Journal,* 14; 51–56.

Appendix E

Greenhouse Gas Inventory Process

Contributed by
Douglas Huxley and Priya Sathyanarayan
(authors of WRF Project 4156, Greenhouse Gas Emission Inventory and Management Strategy Guidelines for Water Utilities.)

UNDERSTANDING THE GHGs OF INTEREST

Based on precedent from the Kyoto Protocol, developed by the United Nations Framework Convention on Climate Change (UNFCCC), and brought into force in February 2005, six greenhouse gases (GHGs) or types of GHGs are typically included in an emissions inventory:

- Carbon dioxide (CO_2)
- Methane (CH_4)
- Nitrous oxide (N_2O)
- Hydrofluorocarbons (HFCs)
- Perfluorocarbons (PFCs)
- Sulfur hexafluoride (SF_6)

Each GHG has a global warming potential (GWP). GWP represents the most accepted measure of the heat-trapping ability of each GHG relative to CO_2. For example, on a unit mass basis, CH_4 has approximately 25 times more ability to trap heat in the atmosphere than CO_2. Hence, the GWP of CH_4, according to the Fourth Assessment Report (AR4) noted below, is 25. Halogenated compounds such as the HFCs and PFCs are much more potent GHGs, primarily due to their long atmospheric lifetime, and thus have very high GWP.

The Intergovermental Panel on Climate Change (IPCC) has published a series of "assessment reports," which include, among other topics, consensus observations of climate changes seen so far and predictions of future changes, analysis of the reasons for those changes, analysis of society's vulnerability to those changes and required adaptation measures, and assessment of options for mitigating changes. The physical science sections of these reports have become the international standard for estimation of GWP, and estimated GWPs have been updated with each report based on changes to sources and sinks of CO_2 and improved scientific understanding. Although AR4 is the most recent report, many GHG reporting programs and registries still require use of Second Assessment Report (SAR) values.

GHG emissions are reported on a carbon dioxide equivalent (CO_2-e) basis, which is the mass of the GHG multiplied by GWP. Thus, 1 ton of CH_4 is equivalent to 21 tons of CO_2-e based on SAR GWP estimates, or 25 tons of CO_2-e based on AR4 GWP estimates.

Registries, Programs, and Protocols

Many international and domestic protocols and quantification tools are available for the voluntary accounting and reporting of GHG emissions at the corporate, facility, or municipal levels. Your utility might use these protocols and tools to generate annual sustainability information releases, report within voluntary reduction programs, or report within mandatory reporting or cap-and-trade programs. While there is no one protocol or tool developed specifically for the use of water utilities to quantify entity-wide GHG emissions, the existing protocols, guidance documents, and quantification tools can be used to develop inventory goals and objectives, organizational and operational boundaries, reduction targets, and an entity-wide GHG emission inventory.

Once you choose a protocol to use to quantify GHGs, your water utility may report the GHG emissions in a GHG registry on a voluntary or mandatory basis. Typically, using a registry will require third-party verification of the GHG emissions inventory for independent assessment of the transparency, rigor, and integrity of the calculations.

The following list of registries, programs, protocols, and tools were referenced as part of the Foundation report:

- World Resources Institute/World Business Council for Sustainable Development (WRI/WBCSD)—The Greenhouse Gas Protocol Initiative
- USEPA—National Greenhouse Gas Inventory
- US Department of Energy—Voluntary Reporting of Greenhouse Gases Program [1605(b)]
- The Climate Registry (TCR)
- The California Climate Action Registry (CCAR)
- International Council for Local Environmental Initiatives (ICLEI)—Local Governments for Sustainability
- Chicago Climate Exchange (CCX)
- International Organization for Standardization (ISO)

- United Kingdom Water Industry Research (UKWIR)—Workbook for Quantifying Greenhouse Gas Emissions

While each of these have relevance to water utility GHG emission inventories, the Greenhouse Gas Protocol Initiative and TCR are two of the most important programs for any reporting entity. These are described briefly below.

The WRI/WBCSD Greenhouse Gas Protocol was the starting point for virtually every protocol that is available today. It consists of two separate but interrelated modules: (1) GHG Protocol—Corporate Accounting and Reporting Standard and (2) GHG Protocol—Project Accounting. The GHG Protocol includes methods of calculation for stationary combustion, indirect emissions, mobile combustion, HFC usage, and sector-specific modules for the cement industry, aluminum production, iron and steel, pulp and paper, ammonia production, etc. However, it does not include a registry for external reporting of emissions. The GHG Protocol standards, calculation tools, and other information can be found at: http://www.ghgprotocol.org.

TCR was developed to set consistent and transparent standards for the measurement, verification, and public reporting of GHG emissions throughout North America in a single registry. TCR is a policy-neutral collaboration currently including 40 US states; 9 Canadian provinces; 6 Mexican states; and 3 Native American tribes, and is aimed at developing and managing a common greenhouse gas emissions reporting system. TCR is a GHG emissions inventory registry and not a mandatory reduction or cap-and-trade program. Registry participants must commit to (1) calculating emissions using the TCR General Reporting Protocol and sector/source-specific guidance and tools provided; (2) hiring an approved third-party contractor to verify their emissions data; and (3) reporting verified, entity-wide GHG emissions data throughout the United States, Canada, and Mexico on the TCR Web site through the Climate Registry Information System (CRIS). TCR protocols and the CRIS application can be found under "Member Resources" at: http://www.theclimateregistry.org.

Accounting Fundamentals

Before beginning an inventory, regardless of which protocol you choose to follow, it is important to understand the overall goal of the inventory and the mechanics of analysis. Important topics to understand include project baseline inventories; what Scope 1, 2, and 3 emissions are; and the definition of an organizational boundary versus a footprint.

High-quality GHG inventories follow five fundamental accounting principles that aid in the development of a sound baseline and subsequent updates. These principles include the following:

- Relevance: ensure that the GHG inventory appropriately reflects the GHG emissions and serves the decision-making needs of users—both internal and external to the organization.
- Completeness: account for and report all GHG emission sources and activities within the defined inventory boundary.

- Consistency: use consistent methodologies to allow for meaningful comparisons of emissions over time. Clearly document any changes to the data, inventory boundary, methods, or any other relevant factors in the time series.
- Transparency: address all relevant issues in a factual and coherent manner based on a clear audit trail. Disclose any relevant assumptions and make appropriate references to the accounting and calculation methodologies and data sources used.
- Accuracy: ensure that the quantification of GHG emissions does not systematically overstate or understate your true emissions, and that uncertainties are reduced as much as practicable. Achieve sufficient accuracy to enable users of your data to make decisions with reasonable assurance of the integrity of the reported information.

Setting a Baseline

Organizations interested in documenting progress towards climate change mitigation goals typically select a base year or baseline against which future emissions are compared. When choosing a base year, you should typically choose the first inventory year in which complete and accurate data are available for estimating emissions across all source categories. Water utilities should be cognizant, however, that external factors may affect whether base year emissions are higher or lower than average year emissions prior to any reduction efforts; for example, selection of a base year with a dry hydrologic cycle may show an artificially low baseline inventory, and following wet years would be inflated by comparison.

In some cases, the reporting program and/or registry under which an entity is reporting dictate the base year. In these instances, the reporting program and/or registry guidelines should be followed.

Scope 1, 2, and 3 Emissions

According to most published protocols, GHG emissions are divided into three categories: Scope 1, 2, and 3 emissions. Scope 1 (direct) GHG emissions are released from sources within the organizational boundaries of your utility. This source category is divided into four subcategories: stationary combustion, mobile combustion, process-related emissions, and fugitive emission sources. Scope 2 (indirect) GHG emissions are released from sources outside of the organizational boundaries of your utility but are a consequence of the utility's energy purchases (for example, purchased electricity: emissions from the power plant that generates the electricity consumed by your utility). Scope 3 (optional indirect) GHG emissions are a broad category that covers all other releases that are an indirect consequence of operations, or which could be within the sphere of influence of your utility. Scope 3 emissions contribute to a utility's GHG footprint, as opposed to "inventory." Examples of relevant water utility Scope 3 emission sources are noted below.

See Figure E.1 for an illustration of the three emission types. A more detailed description of organizational boundary versus footprint is provided in the next section.

Inventory Versus Footprint

Following the process laid out by the above-mentioned reporting protocols and registries, you would define geographic, organizational, and operational boundaries when developing your utility's emission inventory. Geographic boundaries are usually set based on state or national boundaries, when reporting to a particular state or federal program, and may not be particularly relevant to most water utilities. Organizational boundaries are used to define the list of facilities, pipelines, etc., included in the inventory and are based on either ownership or ability to control operations for that infrastructure component. Operational boundaries are used to define the emission sources and particular GHGs emitted from within each facility or component included within the organizational boundaries. Then, at a minimum, all Scope 1 and Scope 2 emissions associated with the defined facilities and sources are included in the emission inventory.

Operation of a water utility may indirectly cause Scope 3 GHG emissions from any other sources from outside of those boundaries. Energy may be spent by customers to heat or transport water, other organizations may use energy to transport water to within your boundaries, energy or process-related GHG emissions are caused by the production of chemicals used in treatment processes,

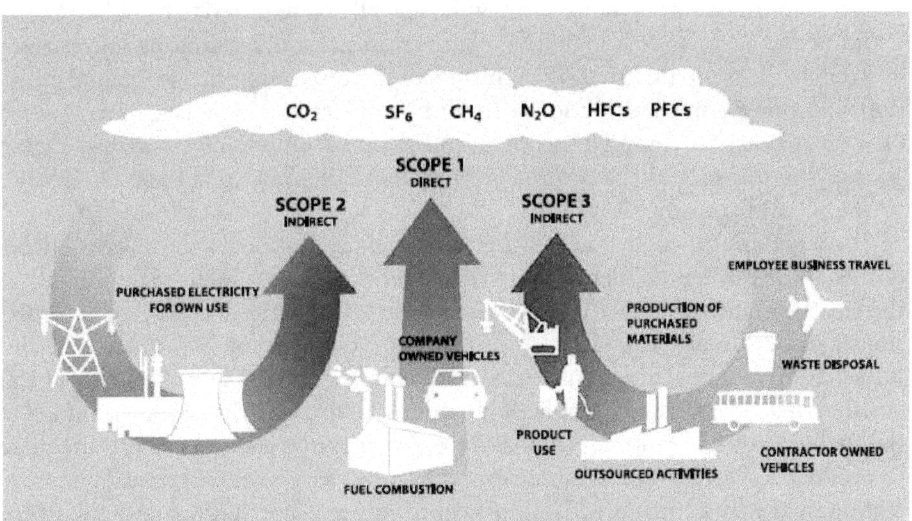

Figure E.1. Three types of carbon emissions (Courtesy of the World Resources Institute)

carbon dioxide is emitted by contracted organizations that supply and regenerate granular activated carbon, and employee business travel and commuting may result in emissions from automobiles, aircraft, or public transportation owned by others. Emissions from suppliers may also be calculated, and other examples certainly exist.

The terms *carbon footprint* or *GHG footprint* are used in several different ways in the climate change management arena. However, it is useful to think of the GHG inventory as the sum of Scope 1 and Scope 2 emissions for all sources within the defined geographical boundaries of the water utility, and carbon or GHG footprint as the sum of the GHG inventory plus relevant Scope 3 emissions.

For water utilities and other organizations interested in reducing the overall climate change impacts of their operations, the GHG footprint certainly provides a more comprehensive view of those impacts and the opportunities available to make a difference. In some cases the most cost-effective and substantial opportunities for water utilities to influence GHG emission reductions may be outside of the organizational boundaries of the utility. Furthermore, many or most water utilities will be exempt from the recently finalized federal GHG reporting rule, proposed federal cap-and-trade programs, and similar state and regional GHG programs. In many ways this leaves water utilities more free to pursue "low hanging fruit" outside of their own boundaries, as compared to regulated entities, which will have significant incentive to pursue reductions of Scope 1 emissions first.

Calculating Emissions

When preparing your inventory, you will typically be required to calculate Scope 1 and Scope 2 emissions, including some emission sources considered *de minimus* (or small, compared to the magnitude of the remaining Scope 1 and 2 emissions). While sample calculations cannot be replicated, below is a general sense of what it takes to complete Scope 1 and Scope 2 emissions estimates. Please consult the full text of the report for sample calculations to aid you during the preparation of your own inventory.

Scope 1 emissions from combustion of fuels in stationary or mobile sources (for example, heaters, boilers, and trucks and other fleet vehicles that are within the entity's organizational boundaries) include CO_2, CH_4, and N_2O. CO_2 emissions can be directly calculated from the total quantity of fuel consumed and information on the carbon content of the fuel. For typical fuels such as natural gas, heating oil, gasoline, or diesel, standard factors for the carbon content of the fuel are available in the protocols referenced above and are usually adequate to produce a relevant and accurate emission inventory, meaning that laboratory testing of the fuel consumed is usually not required. For stationary combustion sources, CH_4 and N_2O emissions are relatively small compared to CO_2, and published emission factors by combustor type are usually adequate. For mobile sources, standard factors are also available for estimation of CH_4 and N_2O emissions based on fuel and emission control type. These factors are reported in quantity of GHG per vehicle mile traveled (VMT), therefore in addition to fuel quantity, VMT must be either tracked or estimated from fuel quantity.

Emission factors for estimation of Scope 2 emissions from use of purchased electricity depend on the mix of electrical generation methods for the power consumed. Different areas of the country have different electrical emission factors for grid power; for example, because of the predominance of hydropower in the Pacific Northwest, a megawatt-hour (MW•h) of power consumed in that region results in lower Scope 2 emissions than a megawatt-hour consumed in a region more dependent on coal power plants. The optimum option is to have your electrical provider give you an emission factor that reflects the blend of types of power used: hydroelectric, gas, coal-fired, etc. However, if that is not possible, grid average emission factors are usually adequate. The USEPA's eGRID database provides regular updates to grid average emission factors based on geographic location. Therefore, for your utility, you can quantify total power consumption by kilowatt-hours and multiply the result by supplier-specific or electric grid subregion-specific emission factors.

In many cases, Scope 3 emissions will involve fuel combustion or electricity consumption, and thus can be calculated with the same methods noted for Scope 1 or Scope 2 emissions. However, Scope 3 emissions can certainly include other emission types (such as process-related emissions from chemical manufacturing), so in some cases more comprehensive knowledge of GHG quantification techniques will be required.

WATER UTILITY ISSUES

Water utilities face some unique issues in implementing an effective GHG management strategy, such as incorporating the concepts of renewable energy and water conservation in a GHG inventory. In addition, it is useful to organize a GHG inventory in a format that is more amenable to benchmarking and later reference. It's good to understand where the bulk of your emissions are being generated so that later, if you implement GHG reduction strategies, you know where to focus your effort. Therefore, we recommend organizing GHG data in six categories: source, treatment, distribution, buildings/infrastructure, fleet, and other.

- **Source.** The source water sector accounts for GHG impacts related to all activities associated with water intake and transport to the treatment facility. All intake structures, pump stations, and screening/solids handling facilities that consume power should be included in this sector.
- **Treatment.** The water treatment facility's power consumption includes that of all treatment processes and the buildings associated with the treatment plant.
- **Distribution.** Every utility will have a unique distribution system that results in a variable use of energy. For many utilities, distribution is the largest energy sector.

- **Buildings/Infrastructure.** This sector collects information on all nontreatment buildings and infrastructure associated with the utility, including all operations and maintenance buildings, administrative buildings, and all other buildings controlled by your utility.
- **Fleet.** This sector includes the energy use associated with use of the vehicles within your utility, including all mobile combustion sources such as cars, trucks, and heavy equipment.
- **Other.** Where possible, all power usage and GHG emissions should be included in one of the five sectors listed above. However, there are times when it is not possible to categorize usage in this manner; as such, this sector can be used for those uses and activities not considered part of the first five sectors.

Water Conservation Issues

The impact of water conservation projects on utility GHG inventories was a key concern during development of the Foundation report. When implementing water conservation, your water utility's energy demand will obviously decrease, and therefore your Scope 2 emissions will decrease. Furthermore, other Scope 3 GHG reductions may result from water conservation, such as power savings from entities conveying water upstream or downstream from the utility, or stationary combustion reductions if the consumer heats less water. But the question arises: who can take credit for these drops in GHG emissions? The water utility? The electric utility? The upstream water supplier that provides water to your water utility? The consumer? In general, this complicated situation does not automatically result in a credit or "offsets" of GHG emissions that the water utility could count as a reduction to Scope 1 or 2 emissions. You should explore discussions with your electrical provider and, if necessary, your upstream water supplier, to determine who can take credit for conservation. A credit can be taken by one entity only, and that will have to be by negotiation. Unfortunately, this is not a clear issue, or a simple way for a utility to decrease its reportable inventory.

Renewable Energy Issues

Another key concern raised by various stakeholders during the development of the Foundation report was in regards to the on-site generation of hydropower. Hydropower and some other renewable power sources result in zero or near zero GHG emissions, and stakeholders requested clarification on when such on-site generation could count as an offset to Scope 1 or 2 emissions. Four typical scenarios exist for the generation of renewable energy by a water utility: (1) the water utility generates green power on-site and consumes it internally; (2) the water utility generates green power on-site and consumes it internally but also sells Renewable Energy Credits (RECs) for the same power; (3) the water utility generates green power on-site, sells it to an electric utility or other entity, and uses grid power for other facilities; and (4) the water utility generates green power on-site, sells it to an electric utility or other entity, sells RECs for the same power, and uses grid power for other facilities. In each scenario, the benefits and permanent reductions to the GHG inventory varied. Please consult the text of the report for a thorough discussion; however, in general, only the first scenario described above would result in zero Scope 2 emissions for the water utility.

MANAGEMENT STRATEGIES

You've completed your utility's GHG emissions inventory and perhaps a footprint analysis, as well. Now that you have a sense of the sources and magnitude of your impacts, you now want to understand how to reduce your emissions. Figure E.2 illustrates one concept for the development and implementation of an overall strategy for climate change response.

In this concept, the strategy and planning begin with an assessment of where you stand now (inventory and footprint analysis) followed by analysis of risks (regulatory, stakeholder initiates, financial impacts, etc.) and opportunities (development of offsets, cost reductions, public relations, etc.). In an iterative process, specific internal and external GHG reduction or sequestration opportunities can be evaluated for costs and benefits, and reduction goals and targets are set based on an assessment of financial and technical constraints. Investment

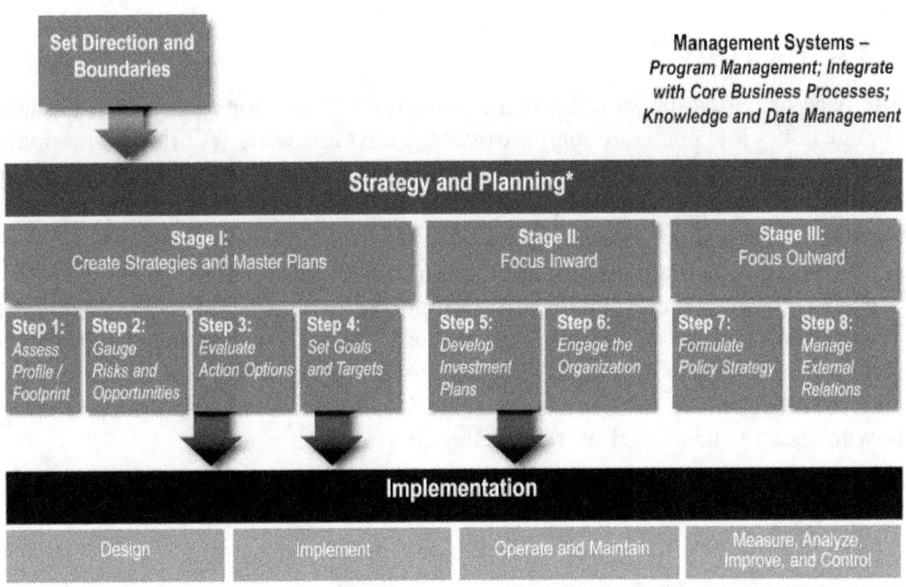

Figure E.2. Plan for developing and implementing a climate change response plan

plans and budget requests are set based on these goals, and engagement of utility management of staff, internal and external communication plans, and policy advocacy would follow. Throughout this process, the design and implementation of reduction projects can proceed; some will be identified as low cost, or having positive ROI, and thus could proceed earlier, while those options requiring larger investment may take longer to develop.

Thus, after the development of an inventory and footprint, you can begin to identify and assess potential internal and external projects. Some examples of internal projects include changes to water supply, energy-efficiency efforts, addition of on-site renewable generation, and fleet management. The options are far broader for external projects. Examples of external projects include water conservation projects, terrestrial or geologic sequestration, community energy-efficiency projects, CH_4 emission reductions, and many other project types that could be pursued through direct funding or via carbon brokers. After your organization identifies potential project types, an overall reduction strategy can be developed, funding appropriated, and external messages developed. The Foundation guidance manual provides further guidance on this topic, including a list of references for energy-efficiency resources.

The process of selecting and implementing a project that is considered an emissions "offset" is one that requires you to consider multiple factors, including the permanence of the project, its relevance, and other facts. Please consult the report for a more thorough description of what it takes to implement an offset in relation to your selected protocol.

Conclusion

The pace of regulatory developments and other drivers for response to climate change issues is expected to increase over the next few years. While we anticipate that water utilities will need to consult program-specific documents and other information sources in development of inventory and management programs, the Foundation report provides the accounting fundamentals and guidance necessary to initiate a management strategy in response to these developments. The concepts we briefly described here are supported by a variety of other subtleties. We suggest that in reading the report, in context of other excellent sustainability resources such as this book, your agency will be successfully on the path toward taking responsibility for your operational GHG emissions and understanding how to lessen your impact on the greater environment.

Appendix F

Elements of Sustainability Coordinator Position

Sustainability coordinator tasks may include those shown below. You do not need to include all of these tasks. These are examples of what others have asked of a sustainability coordinator. Adapt as appropriate for your organization.

- Work with the management team, individual managers, and departments to develop department-specific processes and practices to encourage environmentally sound behavior.
- Form a green team comprised of representatives from each department to plan sustainability efforts, prioritize projects, and coordinate internal efforts with conservation program and other departments to implement best practices relating to sustainability.
- Work with purchasing agent to develop procurement policies that include criteria relating to environmental and social soundness as well as economic viability.
- Work with purchasing agent to research new vendors and new products based on environmental soundness. Assist purchasing agent to develop ways to inform current vendors of your increased focus on sustainability.
- Initiate programs that acknowledge favorable behavior by employees, keep sustainability goals at a high level of awareness, inform the public of successes, and use any other means necessary to accomplish board directives.
- Assist conservation staff with analysis and measurement of success and cost-effectiveness of programs.
- Manage and track agency sustainability goals and metrics.
- Develop systems for making triple bottom line decisions for products and overall business. Work with teams to evaluate decisions based on financial and sustainability goals.

- Manage cross-functional task force team to assist them in developing their key performance indicators and meeting the overall agency sustainability goals.
- Track the task force progress and measure each department's success.
- Compile all metric data to show annual results of goals.
- Prepare annual summary of goals and achievements.
- Promote industry-leading best practices in sustainability. Advise teams on how to find environmental benefits and manage social issues.
- Manage all budgets and savings of sustainability and recommend utilization of funds.
- Assist in managing and developing sustainability communications internally and externally with key stakeholders. Maintain industry research and understanding of sustainability trends, thought leaders, and conferences.
- Create continual educational opportunities on trends and innovations in sustainability field for company employees.
- Manage philanthropic and nongovernmental organization partners to integrate them into the business and develop proactive programs.
- Perform a variety of research work relating to sustainability, including the collection and analysis of various data.
- Attend public meetings of various boards, commissions, and committees, presenting reports on assigned projects and activities.
- May be assigned as a staff liaison providing technical advice and assistance to a specific board, commission, or committee.
- Facilitate intradepartmental and interdepartmental coordination to support agency sustainability efforts.
- Assist in preparing and implementing environmental management systems where appropriate.

Appendix G

Sample Sustainability-related Performance Evaluation Criteria

Samples below are for field and office personnel. Additional criteria sets exist for engineering, fleet, and facilities maintenance.

FIELD CREWS

(Meter Readers, Field Customer Service, Valve, Construction, Maintenance, and Distribution)

These guidelines are a tool in evaluating employee's understanding and adherence to Tualatin Valley Water District's sustainability policies. The guidelines below are educational in nature, illustrating concrete ways for employees to work "smarter" in their place of work. Supervisors may use this list to encourage employees to work on specific points before their next review. Because not all of the guidelines apply to all positions, supervisors are encouraged to tailor them for specific needs.

Note: These evaluation guidelines are new and may need some modification. Please provide suggestions and modifications to Human Resources as they arise.

In the Office:

Source Reduction, Reuse, and Recycle

- Recycles waste paper, beverage containers, plastic, etc.
- Uses nondisposable cups, plates, and utensils
- Makes double-sided copies
- Uses electronic-based communication (e-mail, fax) instead of paper
- Uses fax "Post-it" notes rather than cover sheets
- Reuses paper for notes, phone messages
- Shares memos, newspapers, etc.

- Cancels/reduces unnecessary office subscriptions
- Uses rechargeable batteries in electronic devices such as pagers, meters
- Disposes of used batteries in properly labeled containers
- Turns off lights when leaving office
- Turns off computers when leaving for the day
- Notifies vendors and contractors of TVWD sustainability policies and practices
- Participates in identifying and evaluating additional sustainability policies and practices
- Other ______________________________

Energy/Water Conservation

- Turns off office equipment when not in use
- Uses power-saving modes on computers, copiers, and fax machines
- Turns off lights when not in room
- Uses rechargeable batteries in electronic devices such as pagers, meters, etc.
- Disposes of used batteries and other toxic chemicals in properly labeled containers
- Minimizes water use when cleaning cups, coffee pots
- Other ______________________________

Transportation

- Uses alternative means of transportation (bus, bike, walks, etc.) when possible
- Carpools to meetings
- Uses alternatively fueled vehicles when available
- Encourages guests coming to TVWD to use MAX or carpool
- Other ______________________________

On the Job:

Storage Areas

- Monitors and promptly cleans vehicle fluid spills
- Encourages other employees to maintain a clean vehicle area

Field Operations

- Recycles asphalt and concrete
- Uses recycled product for backfilling excavations
- Uses brooms, sweepers, and vacuums to clean street and alley construction sites to prevent runoff and conserve water
- Encourages other employee participation/cooperation

Sustainability Goals

- Specifies energy-efficient electric motor replacements for pumps, etc.
- Ensures that all scrap materials are recycled

- Uses recycled construction products when possible
- Other ______________________________

Administrative/Office Related Positions

Source Reduction, Reuse, and Recycle

- Recycles waste paper, beverage containers, plastic, printer cartridges, etc.
- Uses nondisposable cups, plates, and utensils
- Makes double-sided copies
- Uses electronic-based communication (e-mail, fax) instead of paper
- Uses fax "Post-it" notes rather than cover sheets
- Reuses paper for notes, phone messages
- Shares memos, newspapers, etc.
- Cancels/reduces unnecessary office subscriptions
- Notifies vendors and contractors of TVWD sustainability policies and practices
- Participates in identifying and evaluating additional sustainability policies and practices
- Other ______________________________

Energy/Water Conservation

- Turns off office equipment when not in use
- Uses power-saving modes on computers, copiers, and fax machines
- Turns off lights when not in room
- Uses rechargeable batteries in electronic devices such as pagers, meters, etc.
- Disposes of used batteries and other toxic chemicals in properly labeled containers
- Minimizes water use when cleaning cups, coffee pots
- Other ______________________________

Transportation

- Uses alternative means of transportation (bus, bike, walks, etc.) when possible
- Carpools to meetings
- Turns in District vehicles for scheduled maintenance
- Encourages guests coming to TVWD to use MAX or carpool
- Other ______________________________

Office Materials and Furnishings

- Specifies recycled content and biodegradable materials, where feasible
- Purchases durable products
- Purchases products with minimal packaging

- Avoids use of office materials that contribute to indoor air quality problems (due to toxic glues/adhesives)
- Requests local goods and services where feasible
- Proactively maintains and services equipment to avoid major purchases
- Other ______________________________

Toxics Use Reduction and Disposal

- Coordinates purchase of potentially toxic products with TVWD Safety Coordinator
- Disposes of used batteries and other toxic chemicals in properly labeled containers
- Other ______________________________

Appendix H

Sample Sustainability Resolutions

RESOLUTION NO. 05-08

A RESOLUTION ADOPTING A SUSTAINABILITY POLICY

WHEREAS, in 2001 the Tualatin Valley Water District (TVWD) embarked on a conscious path to become a "sustainable" water utility, seeking Leadership in Energy & Environmental Design (LEED) certification for our building expansion; examining our policies, products and procedures; and exploring ways to increase our positive effect on society, the environment and the economy, while decreasing our negative impacts; and

WHEREAS, the TVWD Board of Commissioners has initiated and adopted sustainability-related goals for each of the last five years; and

WHEREAS, the District's Board has instructed staff to move forward with sustainable actions such as the District Employee Recycling Benefit, and investigation of solar and micro hydro electricity generation projects; and

WHEREAS, the TVWD Board has thus far accepted two Sustainability Plans detailing additional actions proposed and those accomplished; and

WHEREAS, District staff members have been and continue to be trained in the foundations of The Natural Step and related sustainability basics, and being fully advised,

THEREFORE, BE IT RESOLVED BY THE BOARD OF COMMISSIONERS OF THE TUALATIN VALLEY WATER DISTRICT THAT:

Tualatin Valley Water District will provide quality water and customer service to its community in a manner that protects the earth's environment, supports a strong economy and promotes a robust community. In achieving this vision, TVWD will:

(1) Develop and integrate forward looking, practical and measurable sustainable actions into existing TVWD programs and processes;

(2) Reduce overall cost whenever possible by purchasing materials that are durable and reusable, and when appropriate, incorporate into economic evaluation of materials the environmental and social costs of production, use and disposal of purchased materials;

(3) Work to support and promote the efficient use of water among all customers and employees;

(4) Modify work processes as needed to make more efficient and cost-effective use of materials, equipment and natural resources;

(5) Provide training, information and tools for employees to enable them to provide water services in a sustainable manner and carry those behaviors beyond work to their personal lives;

(6) Provide efficient and effective means to recycle used materials and equipment, and encourage use of alternative energy and increasingly more efficient modes of transportation;

(7) Be a leader in sustainability in the water industry and encourage the development of sustainability policies and practices by others;

(8) Partner with TVWD customers and public, educational, non-profit and private organizations to achieve advances in sustainability.

INTRODUCED AND ADOPTED THIS 16th DAY OF APRIL, 2008
TUALATIN VALLEY WATER DISTRICT

BY______________________ BY______________________

Richard Burke, President Marie Haynes, Secretary

EBMUD

Policy 7.05

EFFECTIVE 10 SEP 08

SUPERSEDES 22 SEP 98

SUSTAINABILITY

IT IS THE POLICY OF THE EAST BAY MUNICIPAL UTILITY DISTRICT TO:

Provide reliable, high-quality drinking water and wastewater service through sustainable operational, maintenance and construction activities that avoid, minimize or mitigate adverse effects to the environment and the public.

Objective

The District will strive to balance environmental, social, and economic objectives into its decision-making, policies, programs, and work practices. In doing so, the District will:

- promote an environmental stewardship ethic in its staff and among other drinking water and wastewater treatment agencies;
- assure that the District adheres to the principles of sustainability and environmental justice;
- comply with environmental laws and regulations;
- look for opportunities for continuous improvement of environmental performance and prevention of pollution to reduce impacts to the environment;
- promote the purchase and use of recycled and recyclable products;
- move towards zero waste and seek ways to recycle materials the District cannot use in its operations;
- establish a framework for setting and reviewing environmental objectives; and
- foster communication with employees and the public about the environmental significance of the District's current and future operations.

Responsibilities

To promote environmental stewardship and facilitate compliance with laws and regulations, the District will conduct compliance audits, administer staff training, and assist in the development and implementation of management and operational practices that support environmental, social, and economic considerations and ensure compliance. The District will maintain strong working relationships with local regulatory agencies, exchanging information on District plans and procedures that support the development of sustainable environmental guidelines for the water and wastewater industry at large.

To advance environmental leadership and awareness, the District will participate in drinking water and wastewater organizations and associations, and work cooperatively with and solicit input from the environmental community and public on District activities.

To promote the use of recycled and recyclable products, the District has a preference for purchasing materials that include recycled and/or recyclable content without compromising the product's fitness, quality, price, and availability.

The District will establish a framework for setting and reviewing long-term sustainability performance objectives. Staff will periodically report to the Board of Directors the status of the District's sustainability efforts which include regulatory compliance, environmental impacts, resource use, stewardship activities, etc.

Sustainability

Sustainability means using resources (economic, environmental, and human) in a responsible manner that meets the needs of today without compromising the ability of future generations to meet the needs of tomorrow. This business approach applies a holistic view and strives to minimize waste; conserve energy and natural resources; promote long-term economic viability; support safety and well-being for employees, communities, and customers; and be beneficial to society.

Environmental Justice

EBMUD will accord the highest respect and value to every individual and community, by developing and conducting our operations, programs, policies, and activities in a manner that promotes equity and affords fair treatment, accessibility, and protection for all people, regardless of race, age, culture, income, or geographic location.

Authority

Resolution No. 32881-94, September 13, 1994
Amended by Board Resolution No. 33120-98, September 22, 1998
Amended by Board Resolution No. 33684-08, September 10, 2008

Reference

Policy 3.02 - California Environmental Quality Act Implementation

---

Appendix I

Sample Sustainable Purchasing Guidelines

I.1. Sustainable Purchasing—TVWD Internal Guidance Document

A. Background

The District recognizes that the purchasing decisions its employees make have important social, human health, environmental, and economic impacts, and that the District should make procurement decisions that embody the District's commitment to sustainability.

In April of 2008, the Board of Commissioners approved Resolution 05-08 adopting a sustainability policy, which is now incorporated into the Administrative Policy Handbook as Section 2.05.110. The Resolution provided that the District was committed to examine its policies, products, and procedures to increase the District's positive impact on society, the environment, and the economy, while decreasing its negative impact. To that end, the District will, among other things:

- Develop and integrate forward-looking, practical, and measurable, sustainable actions into existing TVWD programs and processes; and
- Reduce overall costs whenever possible by purchasing materials that are durable and reusable and, when appropriate, incorporate into economic evaluation of materials the environmental and social costs of production, use, and disposal of purchased materials.

The Board policy is consistent with state law. ORS 184.421 defines sustainability as "using, developing and protecting resources in a manner that enables people to meet current needs and provides that future generations can also meet future needs from the joint perspective of environmental, economic and community objectives."

The purpose of this document is to provide a list of factors or considerations from which to analyze purchases of goods, materials, and products by the District. The District recognizes that the purchasing decisions its employees make may have important community, health, environmental, and economic impacts. The District, when appropriate, should make procurement decisions that embody the District's commitment to sustainability.

In general, District employees should purchase goods, materials, and products in a manner that integrates fiscal responsibility, community, and environmental stewardship. If environmentally preferable or otherwise sustainable items are available, meet the District's specifications, perform satisfactorily, are durable and are priced competitively, then those are the products or goods that should be purchased.

We know that sustainable attributes are not always easy to identify and compare. This document is intended to help you make purchasing decisions or recommendations that include sustainability as one component of your decision. One of the central themes of this document is something the District has considered for years in its purchasing decisions: life-cycle cost assessment. Sustainability is really an enhanced version of life-cycle cost assessment and following that template should get you well down the road in analyzing the type of product or goods to purchase.

If you are purchasing an item that costs more than $5,000 with an expected life of greater than five years, life-cycle cost assessment should be considered.

B. Factors to Consider in Purchasing Sustainably

1. Required specifications

   Will it do what you need it to do? Does it comply with District standards?

2. Quality

   Is it durable? Will it stand up to the conditions under which it is expected to perform? Is there a warranty? Is the vendor reliable?

3. Delivery date

   Can you get it by the time you need it without high shipping costs or special trips to the vendor by a TVWD employee or from the vendor to TVWD?

4. Sustainability

   What impact does this product have on the environment, economy, and community? See Section III for more guidance.

5. Price

   Is this product the lowest priced option that meets the above conditions?

   The type and size of purchase will have an impact on acceptable ranges to determine competitive pricing. If many subjective factors are being considered, an RFP may be the most appropriate procurement vehicle. An RFP should specifically set forth the evaluation criteria and the weighting to be applied so that the District can make a decision as to what procurement is in its best interest. Work with the District Contracts Coordinator to structure an RFP that clearly sets forth these criteria and the evaluation process appropriate for the circumstances. The District has developed a background statement that addresses sustainability. This should be included in each RFP for purchase of goods. When appropriate, include sustainable attributes in product specifications. Please see the following information for further guidance.

C. Determining a Product's Sustainability

Determination of sustainability requires an evaluation of the product's impact on the environment, economy, and community. Though you may not know what sustainable options exist, you should not put anything into a request for proposal or invitation to bid that you cannot measure. If you need to research sustainable opportunities in advance of sending out one of these documents, you have several options:

1. Check the Internet. You can Google anything these days.
2. Ask our current vendors for suggestions. You need to tell our vendors that we are interested in sustainable options or they may not think to suggest them.
3. Ask another staff member for information (TVWD's Strategic Planning Coordinator, Contracts Coordinator, Inventory Buyer, Safety Coordinator, Building and Grounds Supervisor, Field Operations Manager, Engineering Manager, Green Team members, and many others have experience in purchasing sustainable products). These people may also have networks or other organizations, such as the Oregon Public Purchasing Association (OPPA), Public Sector Sustainability Coordinators (PSSC), Partners for a Sustainable Washington County Community (PSWCC), or American Water Works Association (AWWA), that may have already purchased what you are looking for. They can ask their members for assistance and suggestions. The District also uses software designed to identify materials we use that may have negative impacts on environmental and human safety. The soft-

ware also helps us identify alternatives to investigate. The District Safety Coordinator is the primary contact for this information.

4. On larger products, consider putting out a request for information (RFI) to give you an idea of what is available so that you can specify what products you want to get bids or quotes on.

D. Guiding Questions

Keep these questions in mind as you consider purchases. Some of these principles may have applications in other District business decisions beyond purchasing so it may not fit your analysis. But generally, consider:

- Are there alternatives to purchasing a new product? Options include sharing existing products, using surplus products within the District or through the state, or using refurbished products.
- How does this product (or can this product be made to) conserve energy?
- Does this product generate waste? Is there a product that can be used that will generate less waste?
- How is this product packaged? Are there packaging alternatives that can be specified to reduce waste and reduce natural resources used to package the product?
- What natural resources are consumed in the process of manufacturing the product? Are there products that extract fewer natural resources during their manufacturing that should be considered?
- What energy is consumed in the shipping of the product? Are there shipping alternatives that consume less energy?
- Do these products release toxic material during their life cycle? If so what can be done to minimize the release of toxics? Are products available that do not release toxics into the environment?
- What is the life-cycle cost of the product? Would the total cost of the product (operating costs, energy costs, training costs, maintenance costs, licensing, expected life of the product, disposal, and any other costs that can be objectively quantified) result in a different purchasing decision and less cost to the District than if the decision was based only on the initial cost? Refer to the more detailed explanation of Life-Cycle Cost Assessment at the end of this document.
- What has the supplier done to incorporate sustainable practices into their manufacturing and business operations? Are there purchasing choices we can make that will foster sustainability from suppliers?
- Will the purchasing decision anticipated to come from this procurement have a positive impact on the community (e.g., buying locally)?

E. Exporting Our Thinking

There are other ways to encourage sustainable practices. For instance, when interfacing with vendors, conference calls should be requested when appropriate, electronic copies should be requested instead of paper whenever possible, and two-sided copying should be requested when appropriate. Think of what we do in our offices and request that consultants, contractors, etc., follow the same type of practice when dealing with District staff, if appropriate.

F. District Expectations

While not all factors will be incorporated into every purchase, the District expects that employees will make a good faith effort to incorporate and balance these factors to the maximum extent possible. The District also expects that employees will utilize best practices in sustainable purchasing. Best practice and sustainable procurement are those that use the aforementioned sustainability factors, standards, and procedures in an efficient and effective way that is successful and replicable. Opportunities change though, so please check in from time to time to see if there are even better alternatives today than the great one you found last month or last year.

In addition to products, services should also be evaluated for sustainability whenever possible. For instance, conference calls should be requested when appropriate, electronic copies should be requested instead of paper whenever possible, two-sided copying should always be requested. Think of what we do in our offices and request that consultants, contractors, etc., follow the same type of practices when dealing with District staff.

I.2. Signs of a Sustainable Caterer

Ask about some of these actions and items as you evaluate the sustainability of potential caterers.

- Ingredients and products are grown organically and/or locally and are as minimally processed as possible
- Natural, hormone-free meats are used
- Baking is done with organic flour and sugar and free-range eggs
- Shade-grown, fair trade coffee is served
- Buyer frequents farmers' markets or has small, local, organic farmers who supply them with what's best, fresh, and in season
- Offers real plates and cutlery at no extra charge to reduce paper waste
- Uses compostable sugarcane paper products, compostable corn-based plastic hot and cold cups, and/or compostable potato-based plastic utensils when disposables are required

- Uses recycled-content garbage bags
- Uses recycled-content and unbleached napkins
- Uses phosphate-free dish soap
- If available, participates in commercial composting program for both kitchen and post-event composting
- Recycles all paper, cardboard, cans, bottles, and plastic
- Donates leftovers from events to homeless shelter or other similar organization
- Purchases renewable power
- Purchases carbon offsets to offset nonelectrical energy used (e.g., travel)
- Has business cards and brochures printed using soy-based inks and the highest amount of postconsumer recycled material possible
- Uses e-mail whenever possible instead of fax or standard mail
- Purchases company shirts that are locally screened onto sweat-shop-free T-shirts
- Uses a hybrid or biodiesel-fueled delivery vehicle
- Produces trade-show signage on ecoboards rather than foam-core boards
- Has received awards for sustainable performance
- Is part of larger sustainable catering group
- Has pledged to increase their sustainable practices each year
- Has a profit-sharing or performance bonus in place
- Has a green team to evaluate company efforts
- Provides sustainability training for employees

Appendix J

Sample CIP Criteria and Rating Samples

J.1. Tualatin Valley Water District Capital Improvement Plan Sustainability-based Evaluation Criteria

Economy

Supports economic growth and development

- Supports housing and business development in the service area.
- Will provide domestic water and fire flow service levels consistent with comprehensive land-use plans.

Utilization of local employment and manufacturing

- Construction and/or operation of the improvement will result in local employment.
- Local manufacturing of materials and components preferred.

Improves efficiency, effectiveness, reliability; reduces long-term costs

- Reduces system input requirements (energy, water, etc.) per unit of output.
- Improves system throughput or performance.
- Improves reliability of essential system components.
- Results in a reduction of long-term operating costs.

Society

Meets regulatory or contractual requirement

- Provides compliance with state and federal regulatory requirements.
- Meets the District's contractual requirements to others.

Supportive of community esthetics and livability

- Minimizes visual and noise impacts to the community.
- Appearance consistent with community values.

Improves public/employee safety; improves drinking water quality

- Improves public and employee health and safety.
- Improves water quality and esthetics.

Environment

Construction impact on natural environment

- Minimize construction damage to natural systems.
- Postconstruction restoration/improvement of natural systems.

Project sustainability

- Construction and operation input and outputs, to the extent practical, follow sustainability principles (recycled, recyclable, biodegradable, nontoxic, and sustainability harvested).

Supports conservation and/or demand management goals

- Reduces total water consumption of District.
- Influences demand for and usage of a resource or service by adjusting our demand (upward or downward as appropriate) for that resource or service.

Weight and Scoring Guide—Sustainability-Based Evaluation Criteria

Category	Criteria	Weight	Scoring Guidelines	
			Score	Characteristic
Economy	Supports economic growth and development	3	5 3 1	Project will directly allow increased development. Project will only serve existing development. Project incidental to existing and/or new development.
	Utilization of local employment and manufacturing	3	5 3 1	Oregon/Washington employment and manufacturing. US employment and manufacturing. Outside US employment and manufacturing.
	Improves efficiency (1) Improves effectiveness (1) Improves reliability (1) Reduces long-term costs (1) Other economic benefit (1)	4	5 3 1	Five criteria addressed. Three criteria addressed. One criterion addressed.
	Total Category Weight	10		
Society	Meets regulatory or contractual requirement or recognized standards of practice	5	5 3 1	Corrects current violation of regulation, contract, or standards of practice. Prevents future violation of regulation, contract, or standards of practice. Not related to regulation, contract requirements, or standards of practice.
	Supportive of community esthetics and livability	2	5 3 1	Elements of the project enhance esthetics and livability. Project is neutral to esthetics and livability. Project reduces esthetics and livability.
	Improves public/employee safety, including fire protection capacity; improves drinking water quality	3	5 3 1	Exceeds standards for safety/water quality. Meets standards for safety/water quality. Not related to safety/water quality.
	Total Category Weight	10		

Category	Criteria	Weight	Scoring Guidelines	
			Score	Characteristic
Environment	Construction impact on natural environment	3	5	Minimize impact w/ significant improvement of natural systems.
			3	Minimize impact w/ some improvement to natural systems.
			1	Minimize impact w/ restoration as found.
			0	Negative impact on natural environment.
	Project sustainability	3	5	Significant incorporation of sustainability principles.
			3	Modest incorporation of sustainability principles.
			1	Low incorporation of sustainability principles.
	Supports conservation and/or demand management goals	4	5	Significant support.
			3	Modest support.
			1	Project not related to conservation or demand management.
	Total Category Weight	10		

J.2. Lake Oswego CIP Evaluation Criteria: Sustainability Lens

Sustainability means quality of life now and in the future. One way of thinking of sustainability is the three-legged stool: economy, society, and environment. The following worksheet evaluates projects with three equal "legs." It is based on evaluation criteria developed by Tualatin Valley Water District (Ore.), City of Olympia (Wash.) Public Works Department, and City of Lake Oswego (Ore.).

Some categories below may seem unrelated to water source, storage, and distribution projects. The value of looking at capital projects through a sustainability lens, though, is the opportunity to make them even more valuable than they already are. If tennis courts can be constructed on top of a covered storage tank or a wetlands area can be restored as part of a pipe replacement job or it shares resources with others in your locale, you are adding value to your community, the economy, and the environment. This process encourages you to be creative and think holistically.

SCORING AREAS:

Economy

Local economic development

E1. Supports locally owned businesses
E2. Employs local residents
E3. Uses materials produced within 250 miles of work site
E4. Uses materials produced in North America

Economic efficiency

E5. Increases efficiency in an essential system
E6. Increases connectivity in an essential system
E7. Improves reliability of an essential system
E8. Will not increase long-term maintenance/operating costs
E9. Reduces long-term maintenance/operating costs
E10. Sharing resources with others
E11. Inflexible/weds city to a single technology or supplier (minus one)

Society

Social obligations

S1. Assists in compliance with regional, state, federal laws and/or agreements
S2. Is part of an adopted plan

Equity

S3. Supports housing choice/stability
S4. Benefits traditionally underserved populations (e.g., elderly, disabled)

Quality of life

S5. Minimizes visual and noise impacts
S6. Appearance consistent with community values
S7. Improves employee or resident health/safety
S8. Provides alternatives to driving alone
S9. Strengthens social capital
S10. Increases community awareness of sustainability

Environment

Ecosystem functions

N1. Does not increase area of developed land
N2. Promotes efficient use of developed land

N3. Reduces impervious surfaces
N4. Restores natural systems

Project materials and construction methods

N5. Low/zero use of mined materials (petroleum and metals)
N6. Low/zero use of persistent environmental toxins
N7. Low/zero waste produced

Conservation and demand management

N8. Reduces city/resident water consumption
N9. Reduces city/resident fossil fuel consumption
N10. Reduces peak demand on essential systems (e.g., water system)

A project must earn points: the default score for inapplicable criteria is zero, not one. However, all the criteria should apply to almost all the projects. For example, new carpets in a building do not increase the area of developed land, so they DO get a point, even though this is not a "development" project per se.

In general, consider the impacts of a project once it is completed, not the construction impacts. For example, a project gets a point if it minimizes visual and noise impacts or if the final project is quiet and attractive, even if construction process is noisy and ugly.

Additional considerations to differentiate projects of similar rank

- Even distribution of benefits within geographic boundaries of the Agency
- Synergies (e.g., two projects that are done more efficiently together)
- Opportunity costs (e.g., one expensive project versus two cheaper ones of lower rank)

Weight and Scoring Guide—Sustainability-oriented Evaluation Criteria

Project evaluation for: ___________________________________

Economy			
Category	**Criteria**	**Score**	**Scoring Guidelines—one point if "yes"**
Local Economic Development	**E1** Supports locally owned businesses		Project provides a direct benefit to locally owned businesses, e.g., better access to customers, better provision of Agency services to business, direct purchase by Agency.
	E2 Supports local employment		Project will increase opportunities for local employment, e.g., better access to employment centers, increased opportunities for home businesses, direct local employment if job to be done by local contractors.
	E3 Uses materials that are produced within 250 miles of work site		Project's primary material(s) are produced locally and could be sourced locally (recognizing that they may not be, depending on procurement practices). In general: materials in cement are produced locally, asphalt is not, unless it is mostly recycled.
	E4 Uses materials that are produced in US		Project's primary materials are produced in US and could be sourced within US (again recognizing that procurement practices will determine whether they really are). In general: materials in cement are produced in US, asphalt is not, unless it is mostly recycled.
Economic Efficiency	**E5** Increases efficiency in an essential system		Project allows a public system to consume fewer resources such as energy, water, or staff time to produce results.
	E6 Increases connectivity in an essential system		Project completes or connects a portion of the water distribution grid, sewer system, or other essential network.
	E7 Improves reliability of an essential system/structure		Project makes a system such as water distribution more reliable, improves functionality of building that serves the public, e.g., water treatment plant, distribution infrastructure.
	E8 Will not increase long-term maintenance/operating costs		Project results in zero or small increase in maintenance obligations or operating costs.
	E9 Reduces long-term maintenance/operating costs		Project reduces maintenance or operating costs in long term.
	E10 Creates a resource that can be shared between City and another entity		Project increases efficiency by creating a resource that can be shared with other jurisdictions, e.g., other regional water agencies. Often co-funded or grant-funded.
	E11 Proprietary/single technology MINUS ONE		Project may prevent Agency from adopting more efficient/cheaper technologies and approaches as they emerge; relies on proprietary system or single technology.
			TOTAL ECONOMY SCORE (10 POSSIBLE)

Society			
Category	**Criteria**	**Score**	**Scoring Guidelines—one point if "yes"**
Social Obligations	**S1** Provides compliance with regional, state, federal laws		Project moves the Agency towards compliance with regulatory obligations.
	S2 Project is part of an adopted plan less than 10 years old		Project is component of a plan that the Agency is publicly committed to. One point for each plan.
Social Equity	**S3** Supports housing choice/stability		Project supports provision of housing other than large, single-family detached units, or project will support use of existing housing stock rather than tear-downs.
	S4 Benefits traditionally underserved populations		Project will provide special/extra benefits to children, the elderly, or persons with disabilities.
Quality of Life	**S5** Minimizes visual and noise impacts		Project avoids/minimizes sound pollution, is not visually intrusive.
	S6 Appearance consistent with community values		Project is attractive or "fits in" to its location, or is not very visible.
	S7 Improves employee or resident health/safety		Project decreases resident/employee exposure to air pollution, environmental chemicals, noise, driving, or another potential harm, or promotes a healthy behavior (e.g., walking).
	S8 Provides alternatives to driving alone		The project promotes any transit mode other than driving, e.g., walking to school or running errands, biking to work, potential for carpooling, easy access to bus route.
	S9 Builds social capital		The project has the potential to strengthen social networks or community connectedness.
	S10 Increases community awareness of sustainability		The project raises the visibility of the Agency's sustainability efforts or educates the public about environmental or social issues.
			TOTAL SOCIETY SCORE (10 POSSIBLE)

Environment			
Category	**Criteria**	**Score**	**Scoring Guidelines—one point if "yes"**
Ecosystem Functions	**N1** Does not increase area of developed land		Project takes place on land that is already developed with a hard surface (paved, gravel, building) or is landscaped with lawn, irrigated ornamental plants, or ivy.
	N2 Promotes efficient use of developed land		Project promotes more efficient use of already-developed land in the future, e.g., infill, redevelopment.
	N3 Reduces impervious surfaces		Project increases permeability of the ground (removes pavement), does not increase pervious surfaces, or increases pervious surfaces by a small amount only.
	N4 Restores natural systems		Project plants trees, increases total vegetative cover, restores natural areas, or restores stream or wetland functions, improves water quality.
Project Materials and Construction Methods	**N5** Low/zero use of mined materials (petroleum and metals)		Project uses few or no petrochemicals, metals, or minerals in its construction. (If project materials not yet determined, project could feasibly be designed to use few or no mined materials.)
	N6 Low/zero use of persistent environmental toxins		Project uses few or no man-made toxins in its construction. (If project materials not yet determined, project could feasibly be designed to use few or no toxins.)
	N7 Low/zero waste produced		Project does not result in large amounts of landfill waste, e.g., any demo'ed structures can be recycled, plants salvaged, etc. (If construction methods not yet determined, project could feasibly be low or zero waste.)
Conservation and Demand Management	**N8** Reduces City/resident water consumption		Project will reduce City consumption of water from the water treatment plant, or will encourage residents to do so.
	N9 Reduces City/resident fossil fuel consumption		Project will reduce City consumption of fossil fuels, or will encourage residents to do so.
	N10 Reduces peak demand on essential systems (e.g., water system)		Project will help with "peak shaving," e.g., reduce peak system demands for water, electricity, roads, etc. (and thus reduce need for systems expansions to serve peaks).
			TOTAL ENVIRONMENT SCORE (10 POSSIBLE)
			TOTAL SCORE (30 POSSIBLE)

Appendix K

Tualatin Valley Water District Sample Bidder Integrity Policy

On pipe, fitting, and hydrant purchases exceeding $25,000, as part of the determination of bidder responsibility and in addition to all other requirements, the District shall research the OHSA and EPA websites to determine if the apparent low bidder's factory source has a record of OSHA and/or EPA violations. If low bidder's factory has no OSHA or EPA violations, a notice of award shall be made.

If the factory in question has OSHA and/or EPA violations, that bidder shall be contacted in writing with a request to provide the District with a letter outlining how the violation(s) in question are being resolved. Upon receipt of their response, an award will either be given or denied to the apparent low bidder. If the apparent low bidder's award is denied, this process will be applied to second low bidder, and so on.

If an award is denied to an apparent bidder due to unresolved factory OSHA/EPA violations, a letter will be issued outlining the reason(s) for the loss of award to the bidder, the factory and where applicable, and the factory's parent company.

Note that this applies only to open violations. If there are past violations that have been resolved, the vendor is considered to have no violations.

Appendix L

Sample Life-cycle Costing Guidance Document

Environmental factors to be considered include, but are not limited to, the life-cycle costing (LCC) of:

- Pollutant releases
- Toxicity, especially the use of persistent, bioaccumulative, and toxic (PBT) chemicals
- Waste generation
- Carbon or greenhouse gas emissions (including taking into account such things as how far a product must be transported to reach us and that plastics are made from oil, which creates carbon emissions)
- Energy consumption
- Depletion of natural resources
- Impacts on biodiversity

A framework called the *Precautionary Principle* may help when evaluating the comparative toxicity of products and services. It instructs decision makers to take anticipatory and protective measures when an activity raises threats of harm to human health or the environment, even if some cause and effect relationships are not fully established scientifically. Sort of a "better safe than sorry" approach.

When purchasing a substance with a Material Safety and Data Sheet (MSDS) that has not previously been used by the District, please consult the District Safety Coordinator. This person has a short form to be filled out that ensures we have the proper paperwork on file and knows what products we have on site.

Social equity factors to be considered include, but are not limited to:

- Human health impacts
- Use of local businesses

- Use of State of Oregon Certified Minority, Women, and Emerging Small Businesses

You are encouraged to use independent, third-party social and/or environmental (eco) product or service label standards when writing specifications for, or procuring, materials, products, or services, so long as such labels:

- Were developed and awarded by an impartial third party;
- Were developed in a public, transparent, and broad stakeholder process; and
- Represent specific and meaningful leadership criteria for that product or service category.

Whenever possible, use label standards that take into account multiple attributes and life-cycle considerations

If you don't know the answers to the above questions, which is likely, here are some suggestions. Reputable labels include, but are not limited to: Energy Star, WaterSense, Green Seal, and Forest Stewardship Council. Consumer Reports has a Web site that may be helpful in sorting out terms: http://www.greenerchoices.org/eco-labels/eco-home.cfm.

Fiscal factors to be considered include, but are not limited to:

- Use reduction (buy only what you really need)
- Product performance and quality
- Life-cycle cost assessment (LCCA); lowest total lifetime cost
- Leveraging buying power
- Impact on staff time and labor
- Long-term financial/market changes

Appendix M

Advancing the Infrastructure Selection Process: Yarra Valley Water Case Study

Contributed by Francis Pamminger,
Manager Research & Innovation,
Yarra Valley Water,
Mitcham, Victoria, Australia

This 2008 case study presents the work that Yarra Valley Water has undertaken to help it select more sustainable infrastructure options. The work aimed to address perceived barriers to adopting systems different from conventional servicing options. Prior to this work, debate existed as to whether or not alternative servicing options could deliver improvements. The perception was that such options would cost more; require additional material and energy, resulting in increased environmental burden; and increase the business risks. Unaddressed, this uncertainty stopped any future exploration of alternative servicing options.

The case study is presented as part of the National Urban Water Governance Project (NUWGP). The NUWGP was established to understand how we can consciously encourage and support greater uptake of water-sensitive technologies in Australia. It is well acknowledged that in transitioning to more sustainable urban water futures, the urban water sector has been limited by a range of institutional impediments, including fragmented organizational responsibilities, disconnected regulatory and policy frameworks, and prohibitive capital and/or maintenance costs (Brown and Farrelly, 2007; Brown et al., 2007). Previous research with the Facility for Advancing Water Biofiltration (FAWB) highlighted that one of the key enabling context factors for providing resilience to the institutional network for change and learning was demonstration projects (Brown and Clarke, 2007). Additionally, during late 2006, a series of interviews was conducted by the NUWGP in Brisbane, Melbourne, and Perth where respondents were asked to nominate a solution or "way forward" to overcome the litany of barriers identified, to which again, a common response was demonstration projects (Brown et al., 2007).

Why Yarra Valley Water Undertook This Work

Yarra Valley Water is committed to environmental sustainability. It recognizes that the well-being of the community and the strength of the economy are dependent on the health of the environment and accordingly strives to provide water and sewerage services within the carrying capacity of nature.

Yarra Valley Water has a clear and bold commitment to achieve environmental sustainability. It is articulated in a business policy, endorsed by its Board, integrated into day-to-day business activities, and monitored monthly by the Executive through a Balanced Scorecard reporting process. Moreover, environmental sustainability is one of the four key strategic objectives that define Yarra Valley Water's existence. Along with targets for customer service, business culture, and efficiency, Yarra Valley Water aims to "provide services within the carrying capacity of nature and inspire others to do the same."

Getting to this stage was in itself quite a journey. Reflecting on the steps the organization had to take, four major challenges were identified and progressively tackled over a number of years. These included organizational buy-in, an agreed definition of sustainability, awareness of the challenges, and a means of moving from the philosophical to pragmatic (Pamminger and Crawford, 2006).

It is a requirement of the Melbourne retail water companies to apply the "Sustainable Management Principles" in providing their services (Clause 26 in Statement of Obligations, 2008). These principles include water conservation; integration of short- and long-term economic, environmental, and social considerations; and conservation of biodiversity. The application of these principles is an important element in Yarra Valley Water's aim to deliver services in the best way possible.

It is with this background that Yarra Valley Water recognizes that a number of environmental constraints are faced in providing infrastructure services to Melbourne. There is insufficient water to meet future growth. Port Phillip Bay, which already processes about 3,500 tonnes of nitrogen a year discharged from Melbourne's sewage treatment plant, has reached its carrying capacity. Moreover, commonly adopted solutions to generate additional water and reduce nutrient discharge are energy intensive, leading to an increase in greenhouse gas emissions, which in turn exacerbates the very problems we are trying to solve.

Hence Yarra Valley Water considers it a core responsibility to find alternative means to provide infrastructure services in a more environmentally sustainable way.

Key Challenges

Innovators seeking alternative servicing options face significant barriers. The barriers are inherently interwoven in the very processes that have delivered our significant societal improvements, which have come from separating water and sewage and adopting a centralized system. One could argue that this centralized service delivery is a central paradigm embedded in water industry culture. From this perspective, some water practitioners consider alternative decentralized options as a problem. They argue that the environment will be worse off if we use more materials, that costs will increase, and that there will be greater risk to the community.

The site chosen for the research was a typical greenfield site, which in this instance was 3,062 ha (11.8 mi^2) and approximately 40 km (25 mi) north of Melbourne. Over the next 15 years, it is planned to house a population of 85,000. Yarra Valley Water engaged the Royal Melbourne Institute of Technology (RMIT) and Commonwealth Scientific and Industrial Research Organisation (CSIRO) to find the best environmentally friendly infrastructure solution for its new developments, which are split between Greenfield and infill sites (CSIRO and RMIT, 2005). They were specifically asked to help identify ways to quantify environmental impacts, total community costs, and to compare risks between alternative serving options.

Process Adopted to Address Challenges

The methods selected to quantify environmental impacts, community costs, and risk follows.

Environmental Impacts

Prior to this study the general industry view was that selecting a third pipe system in this region would lead to an additional environmental burden. It was argued that the recycled water pipes and corresponding pumps would require additional material and energy to operate; hence, any water savings would be offset by a larger environmental impact.

A life-cycle assessment (LCA) was used to quantify the actual environmental differences between alternative servicing configurations. LCA looks at the environmental impacts of a product or service throughout its entire life cycle. The traditional definition of LCA is "a cradle-to-grave environmental assessment that accounts for all resource use and releases related to the system being studied, and translates this information to the possible harm (or benefit) to the environment and human health" (Curran, 1999).

The LCA methodology can be summarized into the following stages (ISO 14040, 2006):

- Goal and scope definition. Here, the purpose of the LCA is identified and the scope and boundaries of the study are set.
- Inventory analysis. This involves quantifying the environmental inputs and outputs at each stage of the life cycle.
- Impact assessment. The inputs and outputs from the previous step are translated into potential environmental impacts.
- Interpretation. The results hitherto are interpreted with regard to the study objectives.

An LCA of the servicing options identified that all of the alternative options analyzed deliver an environmental benefit when compared against a traditional servicing configuration. Figure M.1 shows the reduction in water consumption, greenhouse gas, and nutrient impacts from each alternative option, compared to traditional servicing.

The options are, respectively:

- **Conventional**—traditional centralized servicing with reticulated water and sewage pipes connected to an existing central network. Water-efficient (3A) appliances.
- **Third pipe from sewage treatment plant (STP)**—third-pipe recycling using treated water from local STP.
- **Third pipe from stormwater**—third-pipe recycling using treated stormwater.
- **On-site water supply and grey water**—water supply solely from rainwater tanks and grey-water treatment at each property.

Community Costs

Prior to this study, the general industry view was that selecting a third-pipe system would cost more than a conventional water and sewerage system, due to increased capital and operating costs for the additional infrastructure.

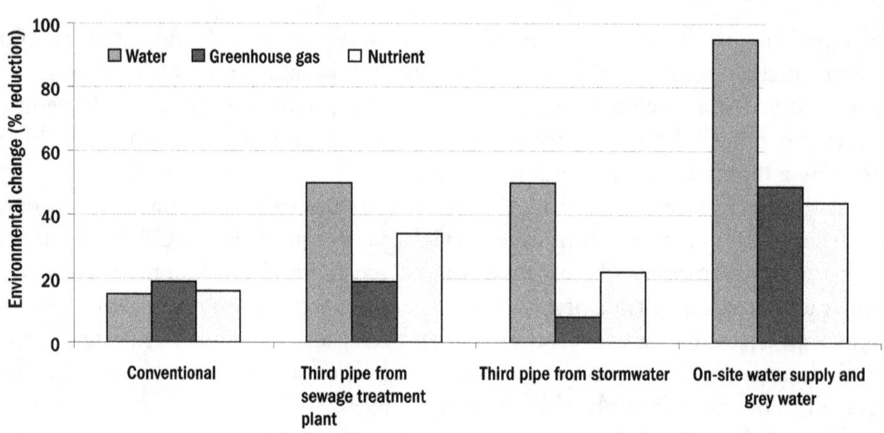

Figure M.1. Life-cycle assessment results

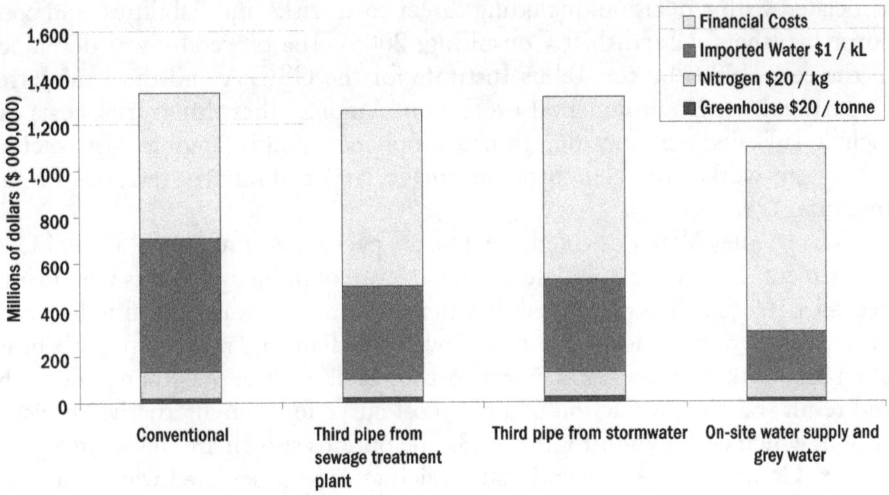

Figure M.2. Community cost results

Life-cycle costing was used to quantify the total costs borne by all sections of the community. This is particularly important with the analysis of decentralized servicing options that can move infrastructure such as rainwater tanks and grey-water systems onto the customer's property. To conduct a true comparison of costs, it is necessary to compare all costs independent of who will be paying for them.

All costs were monetized under this assessment. This included environmental costs such as water, greenhouse gas emissions, and nutrient discharges in addition to the usual capital and operating costs. Water was costed at the market rate of water auctions. Greenhouse gas emissions were monetized at the market abatement rate. Lastly, nutrients were monetized at the rate charged by Melbourne Water in their water quality offsets program for all new developers.

Results from the life-cycle costing showed that when all environmental elements were costed, the decentralized options had a lower community cost. A summary of the results is shown in Figure M.2.

Risk Assessment

Prior to this study, the general industry view was that selecting a third-pipe system would bring additional risks above those occurring with a conventional water and sewerage system. The added complexity of third-pipe systems was considered counterintuitive to the existing adopted industry paradigm to simplify and reduce risk. Third-pipe systems were seen to increase potential for cross-connections, and this in itself was considered a high enough risk to disregard them.

A means to quantify all risks as a cost was required, and a total cost assessment (TCA) methodology was adopted for this purpose. Total cost assessment can be defined as "the consideration of all environmental and health costs associated with a decision, including direct costs, risks and liabilities, and costs borne by others" (Earthshift Consulting, 2008). The procedure was developed in the early 1990s by the Tellus Institute for the USEPA and the New Jersey Department for Environmental Protection. Among other things, risk costs can include lawsuits, penalties due to noncompliance, and intangible costs such as damage to worker morale, corporate image, and community relations (Tellus Institute, 1998).

Yarra Valley Water selected the TCAce package to undertake a Total Cost Assessment. Data were populated using a group of industry experts who identified all potential risks. A probability density function was then fitted to each risk. This was done by identifying the lowest, median, and maximum probability of an event taking place. A business cost for each of these was then added. The end result was the production of a risk–cost curve for each alternative servicing option, which is shown in Figure M.3. The options shown are, respectively:

- **Centralized**—conventional servicing using reticulated water and sewage pipes connected to an existing central network.
- **Dual reticulation**—third-pipe recycling from a local sewage treatment plant.
- **Decentralized**—fully self-contained, utilizing rainwater tanks and package treatment plant (sequencing batch reactor, or SBR) on each property.

More information on the TCAce package can be found at http://www.earthshift.com/tcace.htm.

When risks were costed for a dual reticulation system, decentralized option, and centralized system, it was surprising to find that the centralized system had the highest risk-cost. In essence, this was due to the relatively high probability that drought, fire in the catchments, or algal blooms could occur, and the large number of people that would subsequently be impacted. On the other hand, failures to decentralized systems would impact a relatively small number of people, hence the lower risk-cost.

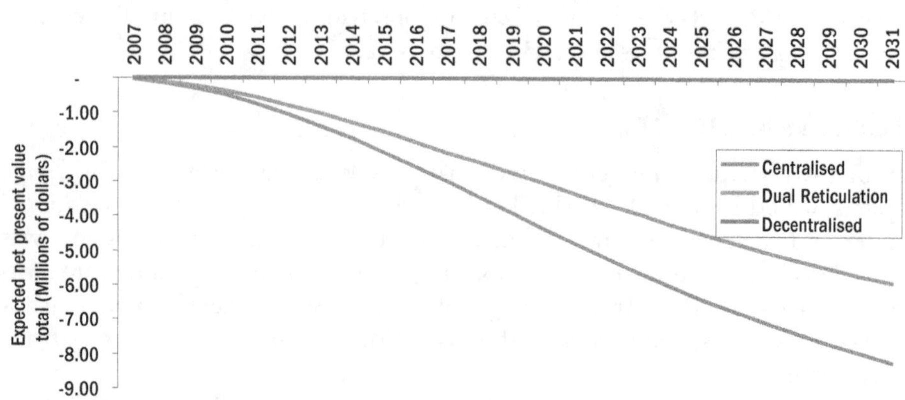

Figure M.3. Total cost assessment results

Conclusion and Lessons Learned

This project analyzed a range of different servicing options for a large 3,062-ha (11.8-mi^2) development in Melbourne. They included both stormwater and sewage recycling, together with on-site options like rainwater tanks and grey-water recycling.

Prior to this work, debate existed as to whether or not alternative servicing options could deliver improvements. The perception was that such options would require additional material and energy use, resulting in an increased environmental burden; would cost more; and would increase the business risk. The use of robust, quantitative assessments has shown that this is not the case for all alternative options in this case study area.

- Life-cycle assessment was used to quantify the environmental impacts, and this revealed that alternative servicing options could deliver a solution with a lower environmental impact than traditional servicing.
- Life-cycle costing was used to analyze the costs borne by all and found that while many options cost more to some individuals, they delivered an overall lower total community cost.
- Total cost assessment allowed a comparative risk assessment of options to be undertaken, and this highlighted that most of the alternative options investigated reduced risks.

The use of life-cycle assessment, life-cycle costing, and total cost assessment has accordingly contributed to changing strongly held views in the industry about the environmental impact of third-pipe systems, community cost, and actual risk.

While it is important to recognize that specific infrastructure options recommended in this study are site-specific, independent on option selection, this work has shown that infrastructure assessments can be made more robust by selecting quantitative methods for critical variables. Critical variables are those variables that most impact the final result.

Introducing scientific methods to an engineering industry, which bases its decisions on scientific principles, has been instrumental in bringing about the change. Barriers to adopting alternative services that previously existed have been removed, allowing a consensus to be reached that will now deliver more sustainable results.

References

Brown, R., and J. Clarke. 2007. *The Transition Towards Water Sensitive Urban Design: The Story of Melbourne, Australia*. Report of the Facility for Advancing Water Biofiltration. Melbourne, Australia, Monash University. (www.urbanwatergovernance.com).

Brown, R., and M. Farrelly. 2007. Institutional Impediments to Advancing Sustainable Urban Water Management: A Typology. In Proceedings of the 13th International Rainwater Catchment Systems Conference and the 5th International Water Sensitive Urban Design Conference, Sydney, Australia.

Brown, R., M. Farrelly, and N. Keath. 2007. *Summary Report: Perceptions of Institutional Drivers and Barriers to Sustainable Urban Water Management in Australia.* Report No. 07/06, National Urban Water Governance Program. Melbourne, Australia: Monash University.

CSIRO and RMIT. 2005. *Sustainability of Alternative Water and Sewerage Servicing Options.* Report produced for Yarra Valley Water.

Curran, M.A. 1999. The Status of LCA in the USA. *International Journal of LCA*, 4:(3); 123–124.

Earthshift Consulting. 2008. http://www.earthshift.com/tcace.htm.

http://www.yvw.com.au/NR/rdonlyres/84699449-D805-4E42-8F3E-A940BCB016E1/0/YVWStageOneFINALREPORT22sept.pdf.

ISO. 2006. ISO 14040: *Environmental Management—Life Cycle Assessment—Principles and Framework.* Geneva: Switzerland: International Organization for Standardization

Pamminger, F., and J. Crawford. 2006. Every Journey Starts with a Single Step: Yarra Valley Water's Journey Towards Environmental Sustainability. In Proceedings of the 12th ANZSYS Conference—Sustaining Our Social and Natural Capital. Katoomba, NSW, Australia.

Tellus Institute. 1998. *Strengthening Corporate Commitment to Pollution Prevention in Illinois: Concepts and Case Studies of Total Cost Assessment.* Waste Management and Research Centre Report prepared for University of Illinois, Urbana–Champaign. http://www.wmrc.uiuc.edu/main_sections/info_services/library_docs/tr/tr30.pdf.

Yarra Valley Water Statement of Obligations. 2008. As amended June 12, 2008.

Appendix N

BAYWORK's Roadmap for Regional Workforce Development

Programs	Challenges	Potential partners	Phase I (1-3 yrs.)	Phase II (3-10 yrs.)
Strategy 1: Get enough of the right people in mission-critical categories				
· Expansion of the candidate pool where needed · Collaborative outreach · Internships · Cross-training	· Insufficient vocational training · Math and communication skills of high school graduates · Inadequate public understanding of water industry opportunities · Inadequate financial resources for utilities alone to fill the skilled labor gap · Accurate quantification of labor needs	· Educational system (K-12, community colleges, higher education) · Professional associations · Department of Labor, Workforce Development Boards · Regulatory agencies · Neighborhood associations · Unions · Other water/wastewater/power utilities	· Refine needs analysis and expand training opportunities for mission-critical skilled trades and engineering classifications · Seek external funding to leverage utility resources · Develop and implement collaborative outreach program	· Use needs assessment processes to identify additional mission-critical classifications at risk and develop programs based on those needs · Expand flexibility of current staff through cross-training
Strategy 2: Give staff the information they need to do quality work				
· Technical training based on competencies needed · Documentation · Knowledge management and access	· Inertia associated with past practices · Energetic creativity required for change · Organizational will and financial constraints · Expertise in information technology	· Professional associations · Other water/wastewater/power utilities · Educational institutions (e.g., community colleges)	· Increased use of technologies such as videoconferencing, video technology, interactive e-learning, blended learning, and knowledge management systems	· Assessment of information and IT equipment needed (e.g., remote technology) to improve information collection and access by water treatment operators

Programs	Challenges	Potential partners	Phase I (1-3 yrs.)	Phase II (3-10 yrs.)
Strategy 3: Modify work processes to optimize use of available staffing				
• Upgraded use of technology • Better match of processes to staff skill-sets	• Time, effort, creativity, and flexibility required to investigate and implement new approaches • Financial constraints	• Other water/wastewater/ power utilities	• Preliminary review of work process improvement opportunities • Preliminary analysis of disconnects between work processes and staff skill-sets	• Information-sharing and potential joint funding of studies in areas of mutual interest
Strategy 4: Maximize cost effectiveness of workforce development investments through collaboration				
• Collaboration with water/wastewater/ power utilities and interested stakeholders at Bay Area, state, regional, and national level • Talent-trading among Bay Area water/wastewater/ power utilities	• Creating new institutional structures and funding arrangements	• Water/wastewater/power utilities • Workforce Development Boards, Dept. of Labor • Educational institutions • Professional associations • Regulatory agencies • Unions	• Establishment of Bay Area Water/Wastewater Workforce Development Collaborative • 2nd West Coast Water/Wastewater Utility Workshop on Workforce Development • Support state and national level efforts • Information-sharing through meetings, workshops, websites, and publications	• Regional collaboration in technical training • Regional collaboration in knowledge management

Index

About the Author

Cheryl Welch earned a bachelor of science degree in economics and mathematics from University of Oregon. After several years as business manager for two arts organizations, she opened a business in Portland, Ore., specializing in environmentally sound paper and printing to fill customers' needs for recycled paper and soy-based inks.

She then combined her financial and sustainability experience and joined Tualatin Valley Water District in 2001, where she developed the district's sustainability program. The program has been honored by such organizations as the American Public Works Association, the City of Portland, Ore., and the City of Beaverton, Ore., for its leadership in sustainability. In 2004, Cheryl was named one of 25 national Environmental Champions by *Interiors and Sources* magazine.

www.ingramcontent.com/pod-product-compliance
Lightning Source LLC
LaVergne TN
LVHW062121200926
848484LV00007B/1734

* 9 7 8 1 5 8 3 2 1 7 9 2 4 *